EARTH DETOX

Every person on our Planet is affected by a worldwide deluge of man-made chemicals and pollutants – many of which have never been tested for safety. Our chemical emissions are *many times larger* than our total greenhouse gas emissions. They are in our food, our water, the air we breathe, our homes and workplaces, the things we use each day. This universal poisoning affects our minds, our bodies, our genes, our grandkids, and all life on Earth. Julian Cribb describes the full scale of the chemical catastrophe we have unleashed. He also maps an empowering and hopeful way forward, to rid our Planet of these toxins and return Earth to the clean, healthy condition which our forebears enjoyed, and our grandchildren should too.

Julian Cribb FRSA FTSE is an author and science communicator. His career includes appointments as scientific editor for *The Australian* newspaper, director of national awareness for the Australian Commonwealth Scientific and Industrial Research Organisation (CSIRO), editor of several newspapers, member of numerous scientific boards and advisory panels, and president of national professional bodies for agricultural journalism and science communication. His published works include over 9000 articles, 3000 science media releases and ten books. He has received thirty-two awards for journalism. His previous books include *Food or War* (2019), *Surviving the 21st Century* (2017) and *The Coming Famine* (2010). As a science writer and a grandparent, Julian is deeply concerned about the existential emergency facing humanity, the mounting scientific evidence for it, and the deficit of clear thinking about how to overcome it. His books seek to map pathways out of our predicament.

D0180505

for those who want a future in which humans can survive. Julian Cribb is one of the world's most important voices on the existential threats of our time.'

Geoffrey Holland, author of The Hydrogen Age

'Julian Cribb refuses to give up on our species. In this his latest book, Earth Detox, he succinctly describes how we find ourselves and our environments contaminated by chemicals while shining a light on pathways to a cleaner, healthier world.'

Robyn Alders, Australian National University

'... harrowing, tenacious and precise analysis of humanity's poisoning of the planet ... From traces of arsenic on the high slopes of Mt. Everest, to flame-retardant chemicals in deepsea squid and bears in the Arctic, pesticides and nanopolluters, to neurotoxins, endocrine disrupters and chemical weapons, Cribb documents the unbearable orgy of toxins we are willy-nilly unleashing ... This is not a book about hope versus despair. It is far more urgent and pragmatic than that. This is about your next breath.'

Michael Charles Tobias, author, filmmaker,
President of Dancing Star Foundation

'... impeccably referenced and scary to say the least. But it needs to be widely read and understood if we seriously want our children to inherit a liveable world.'

Bob Douglas, Commission for the Human Future

'Science policy expert Julian Cribb's latest, deeply-documented, systemic overview of humanity's self-inflicted global crises is essential reading. Cribb provides evidence that our collective expanding cognition and actions offer clear pathways for redirecting societies toward survivable common futures on our endangered planet.'

Hazel Henderson, global futurist, author of
Mapping the Global Transition to the Solar Age,
CEO of Ethical Markets Media, LLC

'Earth Detox is a compelling, albeit unsettling book about the ways billions of tons of toxic chemicals are killing the planet – including people … Julian Cribb takes a no-holds-barred look at how dependence upon manufactured chemicals threatens the world. Every living being is touched in one way or another by toxic chemicals. They're in the air, in the oceans, in the soil, in the foods we eat, on our skin. In fact, almost every facet of our lives is touched by pollutants, whether we know it or not. Earth Detox lays out shocking facts about the many ways we've allowed pollutants to poison the planet. Fortunately, Julian also lays out many actions we can take, and policies to support, to change course and save lives. I encourage readers to pick up Earth Detox, learn the facts about toxic pollutants, and begin sharing some of the solutions with others so we can move toward a safe and healthy world for all beings.'

Suzanne York, Director of Transition Earth

'Confronting, challenging and yet empowering, Earth Detox offers a roadmap to dealing with the issues we all face together as a global community … Julian Cribb reminds us why he is one of the world's leading science writers … If knowledge is power then with this book we are empowered to act, effect change and truly make a difference … At this time of global reflection, the book is guiding light to a better and more sustainable future.'

Ron Ehrlich, President of Australasian College of
Nutritional and Environmental Medicine

'We have treated our planet as an infinitely open system into which we carelessly pour our waste and other toxins. It isn't. Julian Cribb's book is full of well-referenced, irrefutable facts … repulsive in impact, yet captivating, surgically concise and articulate. I commend this book to anyone who cares about our planet and our species.'

John M Schmidt, Synthesist and Foresight Practitioner

Earth Detox

How and Why We Must Clean Up Our Planet

JULIAN CRIBB

CAMBRIDGE
UNIVERSITY PRESS

University Printing House, Cambridge CB2 8BS, United Kingdom

One Liberty Plaza, 20th Floor, New York, NY 10006, USA

477 Williamstown Road, Port Melbourne, VIC 3207, Australia

314–321, 3rd Floor, Plot 3, Splendor Forum, Jasola District Centre, New Delhi – 110025, India

103 Penang Road, #05-06/07, Visioncrest Commercial, Singapore 238467

Cambridge University Press is part of the University of Cambridge.

It furthers the University's mission by disseminating knowledge in the pursuit of education, learning, and research at the highest international levels of excellence.

www.cambridge.org
Information on this title: www.cambridge.org/9781108931083
DOI: 10.1017/9781108946414

First published 2021

Printed in the United Kingdom by TJ Books Limited, Padstow Cornwall

A catalogue record for this publication is available from the British Library.

Library of Congress Cataloging-in-Publication Data
Names: Cribb, Julian, author.
Title: Earth detox : how and why we must clean up our planet / Julian Cribb.
Description: Cambridge ; New York, NY : Cambridge University Press, 2021. |
   Includes bibliographical references and index.
Identifiers: LCCN 2020055149 (print) | LCCN 2020055150 (ebook) | ISBN
   9781108931083 (paperback) | ISBN 9781108946414 (epub)
Subjects: LCSH: Pollution. | Pollution prevention. | Greenhouse gas mitigation. |
   Environmental protection. | Nature–Effect of human beings on.
Classification: LCC TD174 .C75 2021 (print) | LCC TD174 (ebook) | DDC 363.73/6–
   dc23
LC record available at https://lccn.loc.gov/2020055149
LC ebook record available at https://lccn.loc.gov/2020055150

ISBN 978-1-108-93108-3 Paperback

DEDICATION

For Vivienne, Flynn, Adeline and Teddy in the hope they may inherit a brave new world as free of poisons as the one our ancestors enjoyed.

CONTENTS

PREFACE

Earth Detox describes the largest human impact on Planet Earth of all – the hundreds of billions of tonnes of chemicals that we emit and circulate through our normal daily and industrial activities, their impact on human health and on life in general.

It deals with a catastrophic risk that faces all humanity and, potentially, the whole of life on Earth. A curse that is entirely of our making – and which we alone can dispel. Yet one that is not widely acknowledged, in terms of its scale, speed or impact.

Most people are aware that certain chemicals are not good for us and there are too many of these things in our food and environment. Our society recounts numerous 'small picture' stories about chemical accidents, polluted sites, cancer clusters and toxins in food, city air or consumer products.

But these stories are mere scattered pixels in a very much larger picture, one that now directly confronts every living being on our Planet. *Earth Detox* describes the big picture, bringing together peer-reviewed science and evidence from trustworthy international sources. It depicts a menace to our future greater even than climate change, one that will affect every one of us, for all of our lives and for centuries to come. Global poisoning is one of ten catastrophic risks that combine to imperil human survival in the twenty-first century – and like the others, it must be solved and in ways that make none of the other threats worse.

The subject of poison makes for chilling reading, and for that reason the book is short and informative, with plenty of end-notes for those who wish to verify its claims or delve further into its sources. However, no real problem was ever solved or challenge overcome without first grasping its sheer scale, understanding its nature and causes, and developing an action plan. *Earth Detox* lays out both the facts and a credible pathway for achieving this.

Although the topic is grim, this is nonetheless a hopeful book. Whatever we have done to our world, we can undo, and in ways that will lead to better health, greater prosperity and wider opportunities for all. If toxic chemicals are a product of human ingenuity, then ingenuity can overcome their worst effects, leading in turn to innovation, growth, employment and opportunity in the entirely new field of detoxing the Earth. Like renewable energy, the circular economy and renewable food, Planetary clean-up is a cornerstone of our future economic, social and personal wellbeing.

As with climate, this will all depend on an act of willing and enlightened co-operation by people across our world, a voluntary act at species level – maybe the one that ultimately defines us. As with climate, we are all part of the solution.

Read on and be not daunted. Rather, be inspired to help build a better, cleaner, safer world.

1   CHEMICAL AVALANCHE

'There are poisons that blind you, and poisons that open your eyes.'
August Strindberg, *The Ghost Sonata*

The pallid light of a mid-winter afternoon, filtering through a tiny window set high in the wall of the small bathroom, illuminated mother and child in a moment of exquisite tenderness and pathos. Eugene Smith shifted uncomfortably in the cramped chamber to reframe the image: shrapnel wounds sustained in Okinawa as a war correspondent almost thirty years earlier still troubled him. Sightless, deaf, lame, claw-handed and emaciated, Tomoko Uemura lay helplessly in the bath, cradled in her mother's loving arms. Sixteen years earlier she had sustained terrible damage as she still lay in the womb, the venom that crippled her leaching unseen and undetected from the outlet pipe of the nearby chemical plant into the surrounding sea that furnished the food for her village.[1]

Smith was a veteran photographer and photo-journalist who had seen it all – war, suffering, human courage, character and compassion, industry and politics – and depicted it in an epic series of photographic essays, many published in *Life* magazine over several decades. Aroused by growing evidence of the devastation being inflicted on ordinary people by chemical pollution, Smith and his wife Aileen moved in 1971 to the town of Minamata, Japan, following reports of a mysterious disease that had been afflicting its inhabitants since the mid-1950s, to document its impact in images and words. The disease was caused by methylmercury, a substance so poisonous it has no 'safe' level of exposure, no matter how small the dose. It originated in discharges from the local chemical plant. Smith wrote:

> *The nervous system begins to degenerate, to atrophy. First, a tingling and growing numbness of limbs, and lips. Motor functions may become severely disturbed, the speech slurred, the field of vision constricted. In early, extreme, cases victims lapsed into unconsciousness, involuntary movements and often uncontrolled shouting. Autopsies show the brain becomes spongelike as cells are eaten away. It is proven that mercury can penetrate the placenta to reach the fetus, even in apparently healthy mothers.*[2]

The Smiths came for three months. They stayed three years and it almost cost the photographer his life. On 7 January 1972, barely a month after he captured the immortal image of Tomoko and her mother – later to be known as the 'Madonna of Minamata' – he accompanied a group of mercury-poisoning victims to cover a meeting arranged with a manager of the Chisso company which was responsible for running the chemical plant, and thus also for the mercury-laden discharges into local waters where they contaminated the marine food chain on which locals relied. The manager failed to show up. Smith later recounted:

> *But suddenly, a mob of workers rounded a factory building ... They hit. They hit me hardest, among the first. The last exposure, bad, blurred, shows the man on the left, his foot at that moment finishing with my groin, reaching my cameras. The man on the right was aiming for my stomach. Then four men raked me across an upturned chair and thrust me into the hands of six who lifted me and slammed my head into the concrete, outside, the way you would kill a rattlesnake if you had him by the tail.*[3]

Battered and bruised, his cameras smashed, Smith survived but lost partial sight in one eye. It turned out to be his last assignment and he died in 1978.[4]

The bludgeoning of Eugene Smith showed the lengths to which some organisations and individuals were prepared to go to block awareness of the effects of poisons discharged by their enterprises on the community. Despite such attempts to silence the truth, awareness has slowly spread, more so in some societies than others; more in some social strata than others. But the warning has spread neither far nor fast enough: today, most people still have barely an inkling of the universal chemical deluge to which they are now subject, daily, and of the growing peril that we – and all our descendants – face. If the dawn of that awareness for the educated publics of North America and Europe came with the publication of Rachel Carson's powerful book *Silent Spring* in 1962, where she revealed the impact of certain pesticides used in the food chain on wildlife and humans, then Eugene Smith's searing image of the *Madonna of Minamata*, transcending words and languages, was the shot heard round the world.

The subject matter of this book is plain, unvarnished science, as brutal in its facts as the fists and boots that fell on Eugene Smith. But it is the truth, insofar as any system devised by humans is able to determine and describe such things.

***

Earth and all life on it are being saturated with chemicals released by humans, in an event unlike anything that has occurred ever before, in all 4 billion years of our Planet's story. Each moment of our lives, from conception unto death, we are exposed to thousands of substances emitted by our activity, some known to be deadly in even minute doses and most of them unknown in their effects upon our health and wellbeing or upon the natural world. These substances enter our bodies with each breath, with every meal or drink, the things we touch or encounter in our journey through each day. There is no escape from them.

Ours is a poisoned world, its system infused with the substances we deliberately or inadvertently produce in the course of extracting, making, using, burning or discarding the many marvellous products on which modern life depends. Relative to the span of human history, this has all happened quite quickly and has burgeoned so rapidly that most people are still unaware of the extent or scale of the peril in which it places each of us and our grandchildren. Our present plight has crept up on us unseen, piecemeal, with infinite subtlety and frequent inadvertence, in a social climate of trusting acceptance of authority, over barely the span of a single human lifetime. The impacts are only now starting to emerge into full view – and the forming picture portrays a catastrophic risk to our future as great and as all-pervading as climate change, ecological collapse or weapons of mass destruction. A risk to be urgently understood and overcome using all the creative ingenuity humans have relied on for survival throughout our journey.

Knowledge of the toxicity of industrial chemicals is not new: the ancient Greeks and Romans were both familiar with the diseases caused by lead and mercury among those who worked with them, and with silicosis among miners.[5] In the eighteenth century, scrotal cancers were linked with the occupation of chimney sweep.[6] 'Phossy jaw', a disfiguring ailment among workers in matchstick factories in contact with white phosphorus, was first diagnosed in 1839.[7] Aniline dyes, made industrially from coal tar for fabric dyes, poisoned thousands of workers in the mid-nineteenth century – and still do to this day. Poison gases such as chlorine, phosgene and mustard gas – often made by the same factories and firms that produced the textile dyes – inflicted 1.3 million casualties in World War 1. In the twenty years following the discovery of radium by Marie Curie in 1895, a hundred medical workers died from radiation poisoning. Curie herself died, aged sixty-six, on 4 July 1934, of aplastic anaemia, probably caused by prolonged exposure to radiation: she was known to carry test tubes of radium around in the pocket of her lab coat. Asbestos-related cancers and diseases were first diagnosed in the early part of the twentieth century. Following rapid expansion of the coal and petrochemical industries during World War 2, a spate of large-scale industrial poisonings arose: the Great Smog of London, Minamata, Agent Orange, Seveso, the 'Silent Spring', the Love Canal, Bhopal, Dzershinsk, Tianjin, the Asian Brown Cloud and the

Great Pacific Garbage Pool. However, these were mainly viewed by governments and society as single, largely local, misfortunes, the result of corrupt or careless local companies and officials – not as the heralds of a Planetary pandemic.

In our world, something vast has changed.

Today human-emitted chemicals, their byproducts, mixture products and breakdown products, are everywhere, in all that we do. They are to be found in homes, offices and factories; on farms; in clothing, bedding and furnishings; in electronics and plastics; in cars, aircraft and ships; in the air we breathe and the water we drink; in construction and manufacturing industry; in pest control; and in the many products that we put onto or into our own bodies such as cosmetics, medicines, food, drink, tobacco and drugs, both legal and illegal.

Unlike our great-grandparents and all the generations before them, we are now immersed in these human-generated substances 24/7, no matter where we live: the chemical byproducts of modern industrial life have spread around the Planet and their fingerprints are to be found from the remotest poles to the abyssal oceans, from our living blood, to our grave, to the genes of our grandchildren.

In modern society the world over, synthetic chemicals are integral to our daily lives. There is no industry or activity of advanced civilisation where they are not used in some form or other, with the aim of improving our quality of life. They solve problems, protect, adorn, kill pests, save lives, improve efficiency and enhance convenience. An advanced society without such chemicals is almost unimaginable. They are a part of who we are – but in far more ways than most of us suspect. Figure 1.1 summarises the risks associated with common, everyday chemical-based products and services.

In 2018 the United Nations Environment Programme (UNEP) estimated the number of industrial chemicals in general commerce globally was between 40,000 and 60,000.[8] However, the United States Environmental Protection Agency (US EPA) listed more than 86,000 different chemical substances manufactured, used or being researched in the USA alone.[9] (The US chemical

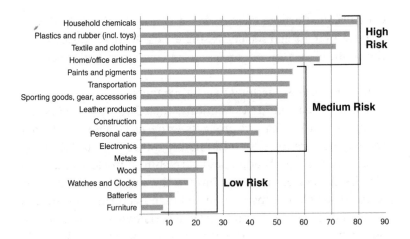

Figure 1.1. Risks associated with common chemical-based products in our daily lives. Source: UNEP.

industry denied this, claiming only 8707 chemicals were used in America.[10]) The US Agency for Toxic Substances and Disease Registry (ATSDR) estimated that 'more than 100,000 chemicals are used by Americans'.[11] The European Union (EU) stated that more than 106,000 chemical substances were used within in its member countries.[12] Furthermore, the EU assessed that almost two-thirds of the chemicals used pose a known health hazard. In China, the Chemical Inspection and Regulation Service (CIRS) listed 49,000 different substances used in the People's Republic, which has become the fastest growing and largest chemical producer in the world.[13] UNEP stated 'The exact number of chemicals on the global market is not known but under the pre-registration requirement of the EU's chemicals regulation, REACH, 143,835 chemical substances have been pre-registered. This is a reasonable guide to the approximate number of chemicals in commerce globally.'[14]

It turns out that all these well-intentioned efforts to quantify the scale of global chemical production were woeful underestimates. In 2020, an international scientific team, led by Zhanyun Wang of Switzerland's Institute of Environmental Engineering, examined the chemical inventories of nineteen countries in Europe, North America, Oceania and part of Asia and concluded:

> *Over 350 000 chemicals and mixtures of chemicals have been regis-*
> *tered for production and use, up to three times as many as previ-*
> *ously estimated and with substantial differences across countries/*
> *regions. A noteworthy finding is that the identities of many chemicals*
> *remain publicly unknown because they are claimed as confidential*
> *(over 50 000) or ambiguously described (up to 70 000). Coordinated*
> *efforts by all stakeholders including scientists from different discip-*
> *lines are urgently needed* ...[15]

The total included 157,000 identifiable chemicals, 75,000 mixtures, polymers and substances of unknown composition, and 120,000 other substances that could not be conclusively identified. Besides revealing that manufactured chemicals far outnumber previous estimates, the world's first-ever attempt to compile a global chemical inventory also pointed to widespread secrecy, mis-identification and obfuscation.

However, even these formal registers cover only chemicals purposefully manufactured by industry. They do not include the far, far larger volumes of substances released directly or as unintended consequences of human activity in the form of construction, land development, farming, mining, mineral refining, energy generation, the use of machinery, deforestation, combustion, transport and other acts that put chemicals in places and concentrations where they would not otherwise naturally occur. And these do not include the millions of breakdown products derived from man-made chemicals, nor the daughter products they give rise to when interacting with other substances in our environment.

Together, these substances – purpose-made and unintentional, simple and evolved – have been entering our lives, our bodies and our living spaces largely unmonitored and, in many cases, undetected, in a rising global flood since the mid-twentieth century, when they first became commonplace. Purpose-made chemicals are thus the mere tip of the iceberg of humanity's total chemical exposure resulting from our own activity.

Without being conscious of it, we poison ourselves every day, every moment of our lives.

Chemical Flood

Chemicals are a burgeoning enterprise – expanding at rates faster, indeed, than economic growth or the human population itself. UNEP anticipates the value of chemical production world-wide to grow from $5 trillion in 2017 to $10 trillion in 2030 and triple again in value by 2050 – a sixfold increase in barely thirty years.[16] The US ATSDR estimates that 'about 1,000 new chemicals are introduced each year'.[17]

The chemical industry is the second largest manufacturing activity in the world. Between 2000 and 2017, its output nearly doubled, from about 1.2 to 2.3 billion tonnes.[18]

Consequently, the UN commented: 'Trends ... suggest that the doubling of the global chemicals market between 2017 and 2030 will increase global chemical releases, exposures, concentrations and adverse health and environmental impacts unless the sound management of chemicals and waste is achieved worldwide.'[19]

Another way to see the issue is that humanity is being exposed to nearly 3 billion *additional* tonnes of man-made substances – including toxins, carcinogens, nerve poisons and hormone disruptors – every year. That's a third of a tonne of man-made chemicals for every child, woman and man on the Planet. This release is effectively cumulative, year on year, and is on track to triple to *one tonne per person* by the mid-century.

To give some idea of the overwhelming scale of this release, during the 'Agent Orange' defoliant campaign (1961–71) in the Vietnam War the total amount of herbicides released for every member of the exposed rural Vietnamese population per year was about 2.5 kilograms; this was subsequently linked to 400,000 dead or maimed and half a million birth deformities.[20] In contrast, citizens globally are now exposed to combined annual emissions of around 325 kilograms of manufactured chemicals *each*.

These numbers are presented purely to give a sense of the scale of the chemical exposure of modern society. No comparison in toxicity is intended, since the overall toxicity of the chemical avalanche is unknown, so many substances never having been properly tested – and almost none of them tested in mixtures. However, while many of these chemicals are deemed harmless in single, small doses – it takes only a tiny quantity of a carcinogen to unleash a cancer – even harmless substances can recombine or break down to form toxic ones. Once used, chemicals never simply vanish. They or their constituents hang around and form new compounds or mixtures, both safe and deadly, in the living environment almost *ad infinitum* – an issue that has become horrifyingly apparent in the particular case of plastics.[21]

Nutrient Cascade

In addition to the 2.5 billion tonnes of manufactured chemicals produced and released each year, humanity also emits vast quantities of nutrients, soil particles, dust, gases and other substances unintentionally, through global agriculture, transport, energy production and manufacturing.

By far the largest part of this category of emissions consists of eroded topsoil, which is released chiefly by agriculture – especially mechanised cropping – forest removal and land clearing for development. Recent estimates of global soil loss range from 36 billion tonnes a year[22] to as high as 75 billion tonnes.[23] Thus, it requires the loss of from 4.5 to 10 tonnes of topsoil every year to feed each of us.

By eating we are now, effectively, devouring our Planet.

While soil is not usually regarded as a 'chemical', nevertheless its release on such a scale has vast biogeochemical impacts on the Earth, on waters and all life, including us. Soil consists of many chemical compounds, some of which are toxic – such as heavy metals or excessively acidic or alkaline minerals which can foul water. Most of these can react with other substances they encounter in the environment, in both the short and long term, creating new products and causing local pollution hotspots or diffuse contamination. Soil loss also leads to dust storms, sedimentation and the silting of rivers, lakes and dams. It is linked with lung disease, allergies, infectious agents such as anthrax and TB and the pollution of drinking water. Above all it causes malnutrition, which leads to many forms of disease and death. On average, human activity is causing the world to lose its precious topsoil at rates from ten to forty times faster than it is naturally replenished.[24] This places the modern industrial food system on a path of no return.

Soil also contains vast quantities of nutrients, both natural and man-made – nitrogen, phosphorus, potash, essential minerals and micronutrients. For example, only 22 per cent of the world's 250 million tonnes of fertiliser made each year actually ends up being consumed as human food;[25] the rest becomes an environmental contaminant that either feeds weeds and algal blooms or pollutes waterways. The staggering volume of nutrients emitted in lost soil ultimately ends up fouling rivers, lakes, groundwater and the sea, where it has created more than 700 'dead zones' in the world's oceans, places stripped of their oxygen and so, largely devoid of life.[26]

Released on such a scale by humans, nutrients have become a dangerous contaminant of the Earth system and now greatly exceed the volumes that circulate naturally in it. Indeed, nitrogen pollution of the Earth's biosphere is now considered to have breached a boundary more perilous, even, than our release of carbon into the atmosphere.[27] A boundary which, in the opinion of the scientists, humanity ought never to transgress for our own safety.[28]

At the same time food production uses around 5 million tonnes a year of specialised poisons designed to control weeds, insects, rodents and moulds in the farming and food chain. Use of pesticides has thus grown tenfold since Rachel Carson warned the world about them in *Silent Spring* in the early 1960s. However, it is estimated that up to 98 per cent of these specialised poisons, generally delivered as sprays, manage to hit a non-target organism, including honeybees, farm workers and consumers.

Agriculture is thus, unintentionally, the source of one of the world's largest acts of general poisoning of the Earth system – an action that, in the absence of far sounder land management, is on track to redouble by the mid-century as demand for industrially produced food rises to meet both a growing population and rising living standards.

Mineral Landslide

The largest release of all is delivered by the worldwide mining industry. In the course of obtaining the 17 billion tonnes of mineral, energy and construction products used by humanity every year,[29] the activities of mining and refining unleash a colossal waste stream many times more voluminous than the metals or energy produced. World mining output has doubled since the mid-1980s and is expected to redouble by the 2050s – along with its waste stream. Only in Europe is it shrinking: in Asia it has grown by 98 per cent and in Oceania by 132 per cent since 2000. The waste it emits has grown in proportion.

The total volume of this colossal waste stream, which consists chiefly of 'overburden' (broken rock, rubble, soil, dust and sediment removed to expose the orebody or coalbed) and 'tailings', crushed material left over after mineral extraction, has never been calculated – and so represents a grave unknown in human contamination of the Earth system. Possibly the gravest of all. The volume of overburden removed to get at the ore or coal varies between one and fifteen times the weight of the product, especially in open pit mines, but averages about three to seven times.

Once metal ores are mined, they must be crushed and refined to extract the desired minerals. This process also generates vast amounts of waste, known as tailings. The ratio of tailings to metal produced varies widely, from 20:1 to 98:1. For example, world production of 60 million tonnes of aluminium also yields

150 million tonnes of 'red mud', the highly caustic residue from bauxite refining.[30] Tailings are customarily deposited in tailings dams or dumps, into rivers and oceans, whence they usually spread through the wider environment over time causing episodes of poisoning of people and wildlife, killing rivers and lakes, and occasionally being released by disastrous dam collapses. There are thought to be more than 18,000 tailings dams worldwide.[31]

No-one has ever compiled an accurate figure for the total wastes from the mining process, but at a conservative estimate it is from three to ten times larger than the useable fraction of metals, minerals and energy products produced – generating between 50 and 170 billion tonnes of mining wastes a year. That such a shocking volume of material should be continually dumped around the Planet each year without even being measured or accounted for contrasts starkly with the effort that now goes into measuring carbon dioxide emissions and their impacts or other forms of industrial pollution. Except at the local level, and in developed countries, the chemical impact of mining contamination on the Earth system has been greatly underestimated, if not ignored altogether.

A possible explanation for this emerges from the work of a group called Earthworks Action, which searched the records of the world's top ten mining companies and established that, together, they dispose of 180 million tonnes of tailings into nearby rivers and lakes, mainly in developing countries where regulatory regimes are often lax or readily corrupted.[32] We may safely assume that the waste-dumping practices of many thousands of smaller mining houses are, in general, far more careless.

Many of the world's 5000+ most contaminated sites listed by the Blacksmith Institute, a New York-based environmental watchdog, are minesites.[33] Individual cases of mining pollution abound in the scientific literature, in government and media reports, and many involve death and disease in local populations as well as ruined landscapes and poisoned waterbodies.

Mining pollution is no trivial matter, either in scale or in persistence. A well-known impact is the acidification of waterways, caused by the chemical interaction of mineral wastes with the environment, especially water. This kills rivers and lakes and renders their water undrinkable. In the case of 'Iron Mountain', California, the world's most famously polluted mine-site, acid runoff was first detected in 1902, continued after the mine closed in 1963, and is expected to last for another 3000 years.[34] Besides acid drainage, mining waste streams may also contain highly poisonous substances such as mercury, arsenic, cyanide, cadmium and lead.

Energy Exposure

To power modern civilisation by conventional means consumes around 7.5 billion tonnes of coal, 4.5 billion tonnes of petroleum, and 4 billion tonnes of gas every year.[35] This vast output of fossil energy is one of the largest human chemical impacts on the Planet – and does not come without grave cost.

Contaminants commonly found in coal include mercury, cadmium, radioactive elements, sulphur and nitrogen compounds, volatile organic carcinogens and other toxins. The carbon released when coal is burned is the major driver of global climate change. The mining and burning of coal also cause acid runoff and acid rain, which damage rivers and forests and are chiefly responsible for acidifying the world's oceans. By volume, the mining, burning and processing of coal constitutes the largest source of toxicity from any human activity. Coal combustion, for instance, releases about 8100 tonnes of highly poisonous mercury into the Earth system every year. This accumulates year after year, affecting the food chain, such as in long-lived fish. Likewise, petroleum contains many substances, such as benzene, that are toxic or may cause cancer.

The burning of all fossil fuels – coal, oil, gas, tar sands etc. – poses multiple threats to human health, especially for developing children and unborn babies.[36] It releases a range of poisons into the environment including polyaromatic hydrocarbons (PAHs), volatile organic compounds (VOCs), mercury, sulphur dioxide, nitrogen oxides and microscopic particles of black

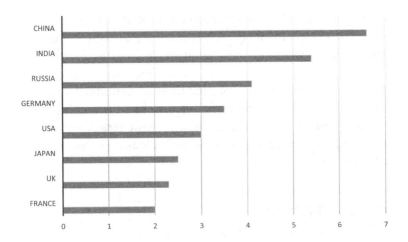

Figure 1.2. Annual cost of air pollution caused by fossil fuels, as percentage of GDP. Source: CRECA.

carbon which impair air quality, water and food. The direct impacts of these on health, especially of infants and children, include lung disease, asthma, developmental and sexual disorders, brain disease and cancers. Furthermore, the burning of fossil fuels releases 37 billion tonnes of carbon into the atmosphere each year, driving climate change, which in turn causes malnutrition, heat stress, infectious disease and mental problems. All of these interact with the direct effects of poisoning to make them worse.

In the world's first endeavour to quantify the costs of air pollution alone, the Centre for Research into Clean Energy and Air (CRCEA) put it at $2.9 trillion in 2018 (Figure 1.2), equal to around 3.3 per cent of world economic turnover.[37] Air pollution also killed around 8 million people from lung disease, heart disease, stroke and cancer, and caused the loss of 1.8 billion days of work, 4 million new cases of child asthma and 2 million premature births.[38]

This, however, is by no means the full toll of fossil fuels, which also includes industrial workers poisoned, the toxic emissions from plastics, packaging, furnishings, paints, cleaning agents and solvents, creating indoor air pollution in homes and workplaces which is often worse than the city air outside. It includes pesticides and microplastics in the home and food chain, the side-effects of drugs both illegal and legal, a flood of 'gender bender' substances and nerve poisons. In short, the fossil fuels industry, while powering society, is also the primary source of most of the poisoning of people and life on Earth.

In October 2017, *The Lancet* Commission on Pollution and Public Health concluded:

> *Pollution is the largest environmental cause of disease and premature death in the world today. Diseases caused by pollution were responsible for an estimated 9 million premature deaths in 2015 — 16% of all deaths worldwide — three times more deaths than from AIDS, tuberculosis, and malaria combined and 15 times more than from all wars and other forms of violence. In the most severely affected countries, pollution-related disease is responsible for more than one death in four.*[39]

The fossil fuels industry, for all its convenience and many economic, social and even health benefits, is thus implicated in the worst case of mass homicide in human history – an annual death toll far greater than that of either of the World Wars.

A toll equivalent in impact to a fresh Holocaust every eight months. A toll that is entirely preventable.

Weaponised Chemicals

Stocks of chemical weapons, continued research into new, more deadly substances and the huge volume of nerve poisons dumped in the world's oceans over the last century pose a continuing threat to human and environmental health. As of 2020, around 70,000 tonnes of the world's declared stocks of 72,000 tonnes of chemical agents and weapons had been declared destroyed under the Chemical Weapons Convention (CWC).[40] In 2020, 193 countries had signed and ratified the convention, while four had not done so. However, cases of the use of chemical weapons in local conflicts and as tools of assassination continued to occur throughout the 2010s,[41] suggesting at least some countries were lying. Concern persists over vast quantities of chemical weapons dumped in the world's oceans between 1918 and 1972, which have never been made safe and may leak out at any future time into the ocean food chain, and over the safety of high-temperature incineration methods used to destroy existing stocks.

Nuclear Legacy

The world produces around 60–70,000 tonnes of uranium oxide every year as the first step in making the fuel for its 450 operational nuclear reactors and for nuclear weapons. The total amount of uranium ore produced globally since mining began a century ago is about 2.5 million tonnes, however the low-level waste generated by the mining, crushing and refining process is thought to run into billions of tonnes. Global uranium reserves total 6.1 million tonnes. In 2019 world stocks of refined product contained 1340 tonnes of highly enriched uranium and 550 tonnes of plutonium. The weapons stockpile held 13,865 warheads, of which 3750 were deployed ready for use.[42] To these sources of radiation must be added the contamination released by nuclear accidents such as Chernobyl and Fukushima,[43] or from disintegrating storage facilities, such as Hanford, USA, where a quarter of a billion litres of radioactive fluids from the H-bomb program are slowly leaching into the environment.[44]

The world's total nuclear waste inventory from power generation, mining, nuclear weapons, decommissioned nuclear plants and nuclear medicine was estimated by the UK at 6.7 million tonnes in 2019 – most of it, low-level waste.[45] The issue of storage remains unsolved: more than seventy years after the start of the nuclear age, no country in the world has a deep geological repository for spent nuclear fuel in operation.[46] There is a fear that, if mounting nuclear waste stockpiles were to be disposed of by illegal dumping at sea, the radiation could affect humanity through the food chain and contamination of the wider environment.[47]

Eating Garbage

The world currently throws away over 10 billion tonnes of domestic and industrial waste materials a year – all of which has chemical consequences for our environment and our health. This mountain of refuse is forecast to almost double by the mid-century. Most of it is stored in city dumps, is burned or simply discarded, mainly in the oceans – all processes that eventually release toxic chemicals back into the air, water and food supply, thus causing us to eat, breathe and drink our own garbage.

Waste from the thirty-seven members of the OECD was estimated at 3.8 million tonnes a year in 2015, and since the OECD represents about half of the world economic activity this implies world total waste emissions of more than 10 billion tonnes in 2021.[48] Reinforcing this, the European Union estimates its total waste output at 2.5 billion tonnes a year.[49] As Europe constitutes 22 per cent of world economic activity, this gives a global waste estimate of more than 11 billion tonnes. All of which has chemical consequences.

Part of this chemical flood consists of 400 million tonnes of 'hazardous wastes'[50] – meaning they are very toxic to the environment and threaten the health of people and animals. Hazardous wastes are the dangerous leftovers from activities such as manufacturing, farming, water treatment, construction, vehicle servicing, laboratories, hospitals, firefighting and other industries. They may be liquid, solid or sludge and contain numerous toxic chemicals, heavy metals, radioactive substances, disease-causing organisms and other harmful materials. Even homes generate hazardous waste when they throw away old batteries, electronic equipment and unwanted paints, cleaning fluids or pesticides. Much of this dangerous flood goes into landfills, whence it leaches into the environment or else is shipped to developing countries where primitive recycling industries ensure the toxins end up recirculating in air, water, food, manufactured products – and people. In 1992 the Basel Convention was signed under which signatory nations agreed to restrict the international transport and dumping of toxic waste.

Every day we throw things away and forget completely about them; but this does not mean these things or their chemical components are gone for good. Many return to haunt us. Chemicals and their byproducts can leach out of landfill pits into groundwater or seep out as vapours. Some of them hitch rides on dust particles, leapfrogging around the Planet in cycles of absorption and re-release in a phenomenon known as the 'grasshopper effect'. Burning of hazardous wastes may give rise to gases and vapours more toxic still. Some break down into relatively harmless substances, but others form more poisonous compounds and many combine and recombine with one another over long periods, giving rise to generations of unintended byproducts. As cities expand, buildings rise over old dumps, their citizens unmindful of what is in them and living on top of potentially toxic zones. In most of the world's big cities today, drinking water contains the lethal residue of yesterday's wastes. While it is possible to clean up one or two toxic substances from a particular site, remediating the witches' brew of toxics produced in the stream of human industrial waste to a safe level is, generally speaking, beyond current technology or affordability, especially once these substances have escaped into groundwater and dispersed through the wider environment.

Finally, the world's water supplies – and through them the entire global food chain – are becoming increasingly contaminated by a wide range of substances capable of damaging the human brain, reproductive and hormonal systems. These are known as endocrine-disrupting chemicals (EDCs), and their main sources include prescribed medical drugs and contraceptives, illegal drugs, pesticides and industrial petrochemicals, plastics and wrapping found in both food and household goods. So far, more than 1500 of these chemicals have come to the attention of science; a few are now banned in Europe, but rarely elsewhere. This issue is discussed more fully in Chapter 3.

To give a fuller impression of the scale of human contamination of the Earth system, Table 1.1 shows the main categories of man-made chemical emissions.

Table 1.1 Global emissions of chemical substances due to human activity, per year

Total disposed waste	10–11 billion tonnes	OECD, EC
Household waste	2 billion tonnes	TWC 2020
E-waste	50 million tonnes	TWC 2020
Hazardous waste	400 million tonnes	TWC 2020
Food waste	730 million tonnes	FSI 2020
Manufactured chemicals	2.5 billion tonnes	UNEP 2019
Pesticides	5 million tonnes	Worldometer 2020
Fertilisers	250 million tonnes	FAO 2016
Plastics	400 million tonnes	WWF 2019
Food output	5 billion tonnes	FAO 2019
Forest output	5.8 billion tonnes	FAO 2018
Total mining output	17 billion tonnes	WMC 2019
Gas	4 billion cubic metres	GESY 2019
Petroleum	4.5 billion tonnes	GESY 2019
Coal	7.5 billion tonnes	GESY 2019
All metals	5 billion tonnes	
Cement	4.1 billion tonnes	Statista 2019
Steel	1.8 billion tonnes	Statista 2019
Uranium oxide	63,000 tonnes	WNA 2019
Mining wastes	36–84 billion tonnes	
	5–10 billion tonnes (tailings)	
Carbon (all sources)	37 billion tonnes	IPCC 2019
Soil, eroded by farming and land development	36–75 billion tonnes	*Nature* 2017
Total emissions per year	**120–220 billion tonnes**	

Chemical Deluge

Table 1.1 represents an attempt to quantify human chemical emissions, in order to visualise the true scale of the problem we are facing, for the purpose of trying to solve it.

Although the numbers are necessarily imprecise, owing to wide variances or lack of data on issues such as land degradation, mining and mineral wastes, it is clear at a glance that our combined emissions of chemically reactive substances are very large indeed, amounting to a billion tonnes every few days.

Total human chemical emissions are thus in the range 15 tonnes to 28 tonnes per person per year – which contrasts with 6 tonnes of climate emissions each. And they are climbing rapidly.

We now inhabit the Anthropocene,[51] an age in which human action occurs at such a scale that it changes the very Earth and its systems, reshaping them beyond repair or recall. An age in which humans have become a geological force.

One of the chief ways we do this is chemical. While this is acknowledged in the particular case of climate, our combined chemical impact is still poorly understood, inadequately researched and quantified and seriously neglected, especially in policies aimed at cleaning up the Planet. Understanding of the problem and attempts to remedy it are therefore occurring piecemeal and without any clear view of the magnitude or complexity of the challenge. And they are failing, disastrously, as this book will show. Like the blind men feeling the elephant in the Indian folk tale, we tend to perceive the parts – but not the whole beast. And this gives us dangerously flawed understanding of its true nature.

It is therefore necessary to give the beast a name – the Anthropogenic Chemical Circulation (ACC), an ugly name for an ugly thing – so that it may be fully seen and understood for the first time, as an essential step in learning how to master it, before it rages amok and ruins us. The ACC describes not only the entire quantum of human emissions of chemically reactive substances, but also the way they interact and mix, travel around the Planet, take new forms, persist sometimes for millennia affecting the health and wellbeing of humans and all other forms of life. The ACC is just like our carbon emissions – only much bigger and far more noxious.

It began with the industrial revolution and widespread burning of coal, but it has expanded massively since the growth of the chemical industries in World Wars 1 and 2, since the emergence of modern industrial agriculture, mining, construction, packaging and pharmaceuticals in the postwar era. Like the radioactive traces of early atomic bomb tests, it is a hallmark of the modern age.

For the first time in the Earth's history, a single species – ourselves – is poisoning an entire Planet.

2 POISONING A PLANET

'Chemical contamination of our environment is the most underrated, under-investigated and poorly understood of all the great risks facing humans in the 21st Century. Its potential impact exceeds even that of climate change.'

Professor Ravi Naidu

Michael Vecchione first went to sea as a cabin boy on a three-masted schooner off Maine, at the age of sixteen. It was the start of a lifelong love of the oceans, their mysterious deeps and all that lives in them which led to his becoming one of the world's leading authorities on cephalopods – squid, octopuses, cuttle-fishes and nautiluses – and Director of the US National Marine Fisheries Service National Systematics Laboratory. Cephalopods, he explains, are essential not only to fisheries management but serve as a general indicator of the health of the oceans and to the future of the twenty-eight species of toothed whales, narwhals, dolphins and other sea creatures that dine on them.

In 2003, Vecchione and colleagues from the US National Oceanic and Atmospheric Administration (NOAA) and Virginia Institute of Marine Science were ploughing through the heaving seas of the western North Atlantic, running a large mid-water trawl between a thousand and three thousand metres (3300–9900 feet) down. In their net they brought up nine different species including short-finned squid, cockatoo squid, jewel squid, vampire squid and the curious jelly-like octopus *Haliphron atlanticus*. Keenly aware of the accumulating pile of research showing that persistent man-made chemicals were being found in whales and dolphins around the world, the researchers decided to examine their catch to see what it might reveal about human impacts in the abyssal oceans.

The results were a shock. These creatures, which had spent their whole lives in the lightless depths hundreds of miles from the nearest human populations, were contaminated with: flame retardants from synthetic furniture and fabrics; cancer-causing electrical transformer chemicals called PCBs (polychlorinated biphenyls) which had been banned for thirty years; traces of the pesticide DDT – also banned – as well as highly toxic anti-fouling chemicals known to cause hormone disruption; also sundry petrochemicals.

'The cephalopod species we analyzed span a wide range of sizes and represent an important component of the oceanic food web,' Vecchione recounted. 'The fact that we detected a variety of pollutants in specimens collected from more than 3000 feet deep is evidence that human-produced chemicals are reaching remote areas of the open ocean, accumulating in prey species, and therefore available to higher levels of marine life. Contamination of the deep-sea food web is happening, and it is a real concern.'[1]

Subsequently, marine scientists have revealed the oceans are awash with toxic man-made chemicals. Dr Tracy Collier of NOAA told the US Congress that man-made industrial chemicals, including flame retardants (PBDEs), plasticisers and surfactants, along with some fungicides and herbicides, are finding their way into fish habitats through surface runoff. The contamination includes historical poisons such as PCBs and DDT as well as synthetic hormones which disrupt the reproduction of sea life.[2]

Local or Global?

Pollution has long been viewed by society, governments, industry and much of the scientific establishment as primarily a local issue. Worldwide, it is conservatively estimated there are more than 10 million potentially contaminated sites,[3] with Europe alone having 3.5 million of these.[4] Yet the term 'contaminated site' implies there is a distinct boundary to the danger zone: that the hazards can somehow be sealed off from the world by putting a chain-link fence around the site of an old factory or dump. However, new research is rapidly accumulating to indicate that the term 'contaminated site' is misleading: that no site is an island, cut off from the rest of its surroundings or from the Planet as a whole, and that most contaminants are, to some extent, perpetually in motion. They are all around us.

Man-made chemicals move constantly in space and time. They travel on the wind, in water, attached to soil, dust and plastic microparticles, in wildlife, in traded goods and in – and on – people. They venture forth as gases and vapours, as liquids and solids, sometimes shifting between the phases. They combine and recombine with one another and with naturally reactive substances to form new compounds – at times more toxic, often less so; some more mobile and others immobile for a time. While some chemicals quickly disintegrate into relatively innocuous components such as carbon, oxygen and water, others seem virtually indestructible as they continually recycle through water, air and soil – and especially in the global food web and the human food chain. The ultra-tough organophosphate pesticides, polyfluoroalkyl substances or PFAS (used in fire-fighting foam) and plastics are examples of persistent substances that hang round for generations, causing harm. Indeed, their persistence is such that 'there is now not a single living organism on the Planet that does not contain DDT'.[5]

Our emitted chemicals do not simply disappear, as many people assume – or as the petrochemical industry prefers us to believe. Their atoms are not destroyed, although their molecular form may alter. They go on forever, reforming, recycling, recombining, reactivating in manifold forms. They become part of an unending, ever-growing, ever-flowing Planetary river: the Anthropogenic Chemical Circulation (Figure 2.1).

In this process, however, what is lost is information about the original substance. It becomes untraceably blended with its chemical environment, both natural and man-made. Manufacturers may be able to trace their particular product along the marketing chain as far as the end user, but rarely

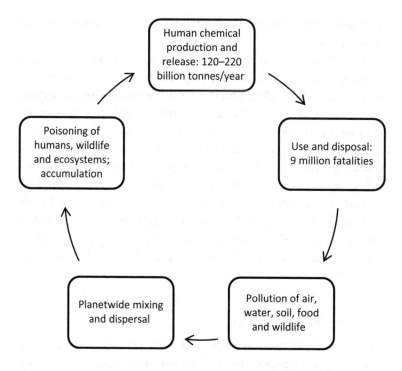

Figure 2.1. Anthropogenic Chemical Circulation. Source: Cribb, 2020.

beyond that and often not even that far. Nor do they choose to inquire too closely. After use, a curtain of oblivion is drawn over human chemical discharges – an opaque curtain that generally only develops tiny holes when a sudden cluster of deaths discloses an untoward exposure in society, as was the case at Minamata, Bhopal, Seveso and the Love Canal.

Humanity is thus increasingly exposed not only to the toxic residues released by our own consumption of material goods, but also those of past generations. And we, in turn, heedlessly inflict our own toxic outpouring on our children and grandchildren – not only directly, but also indirectly, through our damaged genes, as we will see. 'There is already clear evidence that literally no place on Earth, even the most pristine environments that are far from industrial activity, are free of pollutants,' states medical scientist Professor Alfred Poulos, adding, 'pollution is a problem with potentially catastrophic consequences for humankind'.[6]

'Pollution is a grave threat to Planetary health,' Philip Landrigan and nine other leading health scientists warned the world, putting the threat on the same footing as climate, extinction and other well-known catastrophic risks. 'Like climate change biodiversity loss, ocean acidification, desertification, and depletion of the world's fresh-water supply, pollution destabilizes the earth's support systems and endangers the continuing survival of human societies. Pollution, especially pollution caused by industrial emissions, vehicular exhausts, and toxic chemicals, has increased in the past 100 years, with greatest increases reported in rapidly developing low- and middle-income countries. Children are exquisitely vulnerable to pollution,' they cautioned.[7]

Cancer from the Sky

In the early 1970s, scientists realised that almost a hundred man-made substances, chiefly chlorine- and bromine-based compounds used to cool refrigerators and air conditioners and as spray-can propellants, were destroying the ozone layer in the upper atmosphere – the gaseous layer that shields life on Earth against the bombardment of deadly ultraviolet-B (UVB) radiation in sunlight. UVB is a primary cause of skin cancers, eye diseases and wrinkly aged skin. Research on these ozone depleters, known as CFCs, gave us irrefutable proof that even tiny quantities of man-made substances (relative to the atmosphere) can affect the entire Planet, posing a risk to everyone as well as to many other life forms. Thus, using a hairspray in America or France could contribute to a melanoma victim's painful death in Australia by accelerating the destruction of the protective ozone layer over Antarctica that once shielded people in the southern hemisphere from UVB rays, when they worked or played in the sunny outdoors.

Action to end the use of such ozone destroyers began with the signing of the Montreal Protocol on Substances that Deplete the

Ozone Layer in 1987 by 196 nations and the EC.[8] It remains the only treaty ratified by every country on Earth. A third of a century on, despite a thriving global blackmarket in dirty old refrigerators and air conditioners and the durability of CFC molecules themselves, 98 per cent of CFCs had been phased out of production and the ozone hole over Antarctica has started to heal – though that will take another thirty years. This achievement led to the Montreal Protocol being hailed as the most successful environmental treaty ever.[9]

The flip-side to the success story of the Protocol was the fact that the original CFCs were largely replaced by new chemicals, hydrochlorofluorocarbons (HCFCs), which – despite their ozone-sparing qualities – turned out to be highly active greenhouse gases, some of them having more than 2000 times the potency of carbon dioxide, contributing an extra 0.4 billion tonnes of climate heating into the atmosphere each year. The Protocol was then amended to deal with this new threat, with a target of 2030 for total phase-out.

As is so often the case with chemicals, humanity has traded one risk which was fairly well understood for another, equally bad, but still largely unknown at the time the new chemical was released. This lack of precaution prior to release is embedded in industry practice. Chemists design new substances, but frequently avoid asking – or are not resourced to ask – whether they may cause harm to people, infants, wildlife or life-support systems such as the atmosphere, rivers and farmers' soils.

Nonetheless, the successful elements in the story of the ozone depleters and the Montreal Protocol proves that humanity is capable of taking unified action on a catastrophic risk. It is also an object lesson in how new 'solutions' can cause just as much harm as the original threat, if they are not carefully tested first and preventative measures taken.

Every Breath You Take

Air pollution has been a matter of civic concern since at least 1271, when England's King Edward I banned the burning of sea-coal in his capital, London, because of the smog it was causing. Indeed, the citizens of ancient Rome frequently grumbled about the *gravioris caeli* (heavy atmosphere) that hung over their cities from potteries, furnaces and wood-burning domestic fires. What is relatively new, however, is the appreciation that air pollution is no longer confined to individual cities or industrial basins but now flows freely, like a gigantic river, around the globe – and that what is produced in India or China often ends up in the lungs of individuals in Canada or Germany, and vice versa, with harmful effect.

Global air pollution caused by unconstrained industrial development, especially in the emerging economies – as well as activities such as forest burning and transport – has multiplied in line with the human population and our burgeoning demand for energy, goods and services. Air pollution has widely documented impacts on our health. Japanese researcher Hajime Akimoto recounts: 'When the first measurements of high concentrations of CO [carbon monoxide] over tropical Asia, Africa, and South America ... were made available in 1981 [from instruments located] on the space shuttle Columbia, it became clear that air pollution was an international issue.'[10] In his ground-breaking report he went on to explain that global air pollution by ozone was now jeopardising agriculture and natural vegetation worldwide and having a strong effect on climate. During the 1990s, nitrogen oxide emissions from Asia overtook those of North America and Europe, and will continue to exceed them for decades to come.

Air pollution now moves freely around both the Earth's hemispheres and across its largest oceans and continents, obscuring our view of sun, moon and stars from many regions and producing ugly phenomena like the 'brown cloud', the vast grey-brown stain that hangs in the skies over much of southern Asia and consists mainly of soot particles and other pollution from sources such as wood stoves, forest fires, diesel vehicles, factories and coal-fired power plants.[11] As *The Economist* vividly described:

The fetid smog that settled on Beijing in January 2013 could join the ranks of ... game-changing environmental disruptions. For several weeks the air was worse than in an airport smoking lounge. A swathe

*of warm air in the atmosphere settled over the Chinese capital like a
duvet and trapped beneath it pollution from the region's 200 coal-
fired power plants and 5m[illion] cars. The concentration of particles
with a diameter of 2.5 microns or less, hit 900 parts per million — 40
times the level the World Health Organization deems safe. You could
smell, taste and choke on it. Public concern exploded. China's hyper-
active microblogs logged 2.5m[illion] posts on 'smog' in January
alone. The dean of a business school said thousands of Chinese and
expatriate businessmen were packing their bags because of the
pollution.*[12]

Air pollution kills, on average, 8 million people a year, according
to the World Health Organization.[13] It is the fourth most
common cause of death in humans and is responsible for more
than a tenth of global mortality.[14] It afflicts an estimated 95 per
cent of the world's population and has been described by the
WHO as a 'global public health emergency'.[15] In children and
adults, short-term and long-term exposure to dirty local air can
lead to reduced lung function, respiratory infections, aggravated
asthma, and loss of intelligence equivalent to losing a full year of
schooling.[16] 'High air pollution can potentially be associated
with oxidative stress, neuroinflammation, and neurodegeneration of humans,' explains Derrick Ho, a researcher at Hong Kong
Polytechnic University.[17] The current state of the world's outdoor air can be viewed on a real-time satellite map compiled by
the World Air Quality Index, a global collaboration of environmental agencies, on https://aqicn.org/.

Deaths from polluted air mainly result from heart disease,
stroke, chronic obstructive pulmonary disease (COPD), lung
cancer and acute respiratory infections. It increases death rates
from infections such as the pandemic coronavirus.
Contamination consists largely of minute particles, ozone (O_3),
nitrogen dioxide (NO_2), sulphur dioxide (SO_2) and assorted
hydrocarbons which mostly originate with the burning of fossil
fuels in power stations, industrial emissions, landfills, inciner-
ators and transport in all its main forms. In addition to their
climate impacts, fossil fuels thus account for the lion's share,

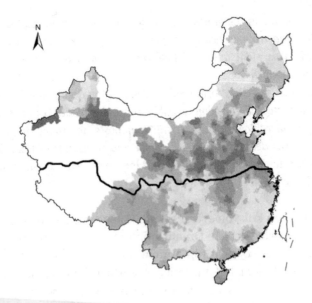

Figure 2.2. The Huai River divides China in terms of coal use, air pollution – and 3.1 years of life expectancy. Shading indicates pollution levels. Source: PNAS.

8 million of the 9 million deaths caused by total 'environmental pollution'.[18]

In the 1950s, the Chinese Government decided it was going to subsidise the burning of coal for home heating north of the Huai River, which neatly divides the country into the cool northern and warmer southern regions (Figure 2.2). The result, several decades on, is that northern China's air quality is notably worse than that in southern China and – significantly – people living north of the Huai enjoy 3.1 years less life, on average, than do people in the south.[19]

Nowadays, indoor air pollution is as serious a problem as outdoor air pollution, being responsible for 4.3 million of the 8 million air pollution fatalities,[20] with the added danger that it cannot be seen by its victims and is therefore often under-rated or ignored. Indoor air pollution has two main sources: the use of wood, coal or kerosene cooking fires in the developing world and the toxic fumes emitted by things found in most homes and cars in the developed or newly industrialised world: furniture, wall and floor coverings, building materials, paints, plastics, foam rubber, bedding, pesticides, cleaning agents and other oil-based products. The poisonous volatile vapours emitted by these items ensure that babies are contaminated from the moment they enter the world, according to the British Parliament. 'Mums in the UK have some of the world's highest concentrations of flame retardants in their breast milk, some of which have now been banned. Chemical flame retardants are still being widely used in our furnishings from children's mattresses to sofas,' said Mary Creagh, chair of the House of Commons Environmental Audit Committee.[21]

And yes, that lovely 'new car' smell is poisonous. In 2012, a team led by the USA-based Ecology Center's Jeff Gearhart tested 200 of the most popular new vehicles for substances that off-gas from parts such as the steering wheel, dashboard, armrests, panelling and seats. They found a noxious cocktail including bromine (associated with flame retardants), chlorine (indicating PVC and plasticisers), lead and heavy metals. These chemicals are linked to health problems such as allergies, birth defects, impaired learning, liver disease and cancer. 'Automobiles function as chemical reactors, creating one of the most hazardous environments we spend time in,' Gearhart stated.[22]

Both forms of indoor pollution – domestic fires and toxic fumes – are linked by medical science to cancers and crippling lung diseases. Modern urban citizens spend between 90 and 95 per cent of their time indoors; thus, their exposure to polluted indoor air is higher and far more prolonged than to outdoor pollution. The irony is that opinion polls consistently show the public is most concerned about outdoor air pollution – presumably because it is visible, whereas indoor pollution is not. Thus, many people seek refuge from outdoor smog by going indoors – where the invisible miasma may in fact be more dangerous to their health. Furthermore, and perhaps for the same reason, there is in general far less government action to control indoor air pollution than outdoor, and to properly regulate the industries that manufacture risky products.

Polluted Poles

The polar bears that roam the Arctic would seem to inhabit one of the purest places on Earth, far from the grime of urban industrial society. Not so. An international team of researchers reported finding the same flame-retardant chemicals as those found in deep-sea squid and mother's milk, in adult Arctic bears and their offspring, studied over a period of twenty-seven years, leading to adverse impacts on the bears' health.[23] The levels in the animals' bloodstreams had more than doubled during the quarter-century of monitoring – yet wild bears are not directly exposed to toxic vapours from modern indoor furnishings and synthetic fabrics, so how did they become contaminated? Scientists consider the answer lies in the escape and Earthwide propagation of these substances in air and water from highly polluted urban environments, then their precipitation in rain and snow, or their washing down rivers and concentrating in fish and other animals in the bears' (and our own) food chain. Similarly, Italian researchers described the finding of significant amounts of industrial mercury, mainly sourced from coal burning in Europe and Asia, amid the Arctic snows, calculating that some 270 tonnes of the highly poisonous metal showers over the Arctic yearly.[24]

At the opposite end of the globe, in Antarctica, the impact is even more striking, as the main source of pollution lies in the industrial heartlands of the northern hemisphere – yet, unexpectedly, the northern filth has somehow crept as far south as is possible on Planet Earth. Australia's Antarctic Division states: 'Minute traces of man-made chemicals used in other parts of the world are now being detected in the snow that falls over the region. Some of these chemicals can become concentrated in the bodies of local wildlife, such as seals, penguins and whales, and can be harmful to these animals in the long-term.'[25] A 2008 review by Simonetta Corsolini concluded 'concentrations are low with respect to other regions of the world, although in some specimens/species (e.g. leopard seal, some invertebrates) they are occasionally high and comparable to those found in regions with a strong human impact'.[26]

In light of the destruction of the Antarctic ozone layer by chemicals manufactured in the northern hemisphere these discoveries ought not to surprise us, yet it still comes as a shock to learn that the most highly protected region on Earth is unavoidably besmirched. Pollution in the Antarctic came to global scientific attention in the late 1970s, following which a long series of studies gave rise to growing concern. In 2009, the International Council for Science reported that consistent levels of persistent organic pollutants (POPs), including pesticides, had been found in the atmosphere, seawater, ice, snow, marine environment and wildlife of Antarctica.[27] By backtracking, the scientists concluded these pollutants had originated as far afield as northern Europe and Russia and had probably mostly reached the southern polar regions through the air. Significant levels of PCBs and hydrocarbon pollution were found in the surface waters of the Ross Sea, which lies off Antarctica south of New Zealand. Industrial chemicals were also found in krill, plankton, fish, penguins, birds, seals, whales and killer whales. These were generally found at lower levels than in similar animals living closer to human populations, but were widespread in the Antarctic wildlife and environment nonetheless. A recent study disclosed significant levels of pollution by heavy metals, with potential to affect algae at the bottom of the food chain.[28] And a third found extensive pollution by plastic particles in Antarctic penguins, at the top of the food chain.[29] Plastics, it should be noted, are one way that other noxious chemicals hitch a ride around the Planet.

Antarctica is most threatened by another major industrial emission – carbon dioxide. Global warming, driven by 42 billion tonnes of human carbon emissions annually, is causing catastrophic changes in polar ecosystems as well as mass melting of the icecaps, on a scale that threatens eventually to drown most of the world's large cities. Antarctica alone carries enough ice to raise world sea levels by 58 metres,[30] and along with the icecaps of Greenland and various high mountain chains, the longer-term potential is for a rise of 66 metres (216 feet).[31] A world team of polar scientists used satellites and modelling to estimate that, between 1972 and 2019, the great southern continent had already shed around 2.7 trillion tonnes of ice, raising sea levels by 7.6 mm.[32] The point about sea level rise is that it is almost unstoppable, unless the Earth chills sharply to pre-industrial temperatures: this means those coastal cities will inevitably drown, even if it takes several centuries. Their condition is already mortal, although most of their inhabitants do not yet realise it.

The Earth's highest point, Mount Everest, is sometimes called the 'Third Pole' by virtue of its altitude and remoteness. In 2006, American scientist Bill Yeo ascended the mountain's Rongbuk Glacier and clambered up its Northeast Ridge as high as 7752 metres (about 1100 metres below the summit), collecting samples of soil and freshly fallen snow for analysis. The results were an unpleasant surprise – the snow held traces of arsenic that were above the safety limit set by the US Environmental Protection Agency (US EPA) for drinking water, with the levels rising with altitude. The samples were also found to contain risky levels of cadmium. While the origin of these poisonous substances is not entirely clear, their levels were higher than in the local rock and three to four times higher in Everest snow than in fresh snow from Antarctica, causing the scientists to suspect Asian industrial air pollution and dust from degraded farm landscapes as the most likely local pollution sources.[33]

The industrial contamination of Mount Everest was not only a shock to mountaineers – who drink the melted snow – but also holds serious implications for millions of people around the world. High, icy mountains such as the Himalaya, Hindu Kush, Rockies, Andes, Urals and other great chains hold one of the largest stores of clean, fresh water on the Planet – the so-called 'water tower': each year their snows and glaciers melt, shooting vast volumes of water down the rivers onto the plains below where people use it for drinking, washing and to grow food. No law protects people anywhere against the insidious pollution of snow melt.

In addition the Himalaya is also the source of one of the worst cases of mass poisoning in human history – affecting an estimated 20 million people living in Bangladesh and Indian East Bengal, from arsenic in their drinking water, which is drawn from wells drilled in their homes and villages by well-meaning aid agencies.[34] The arsenic originates in the rocks of the Himalayan mountains, eroded over time to form the soils of its foothills.[35] However, with the post-World War 2 population boom on the subcontinent, massive land clearing, overgrazing and overcultivation now release a billion tonnes of arsenic-rich topsoil a year, which floods down the region's mighty rivers during the monsoon and is deposited to form the ever-growing Ganges delta, where millions live and draw their drinking water. The poisoning of the Gangetic people is therefore, in part, the result of destructive food production practices and deforestation taking place a thousand kilometres distant, higher up the catchment, adding to the natural erosive cycle. It is a clear instance of the toxic impact that even innocent human activities, such as food production, may have on people living far away.

Distant Domains

For several decades, scientists have been recording disturbing traces of industrial pollution in marine animals such as whales, dolphins, fish and seabirds, often sampled many thousands of kilometres from urban centres of pollution. For instance, in 1997 Heidi Auman and colleagues reported finding highly toxic PCBs, as well as traces of the pesticide DDT, in albatrosses – chicks, adult birds and even eggs – on isolated Midway Atoll in the heart of the Pacific Ocean.[36] The levels they found were comparable with those measured in water birds on the American Great Lakes, which are surrounded by dirty industry: levels close to the concentrations that cause chicks to die in the egg. At the time it was thought this might be due to the dumping of industrial waste into the Pacific, and while this is likely, it is now thought these highly toxic substances also circulate in water, wind, rainfall and the food web, eventually reaching even the remotest places on Earth.

In 2002, Derek Muir of Quebec University and colleagues reported much higher levels of pesticide residues: up to 60 times higher in albatrosses from the north Pacific than from the south Pacific, the difference being due to the fact that the northern bird population fed off the American coast, close to concentrations of agriculture and population, whereas the southern birds fed around New Zealand and in the remote, cleaner Southern Ocean. Nonetheless, all the birds were found to be polluted to some degree.[37]

A similar gradient in pollution was observed in the early 2000s in harbour seal pups, by US and Canadian scientists who sampled along the coasts of British Columbia and Washington State from remote Queen Charlotte Strait to moderately industrialised Georgia Strait, and then to heavily industrialised Puget Sound. The pups from Puget were 'heavily contaminated' with PCBs and other organic chemicals, while those from the more remote places were less so, causing the scientists to suggest that harbour seals might be a useful 'canary in the coal mine' for the effects of heavy industry on the seas.[38] The seals also demonstrate the self-evident fact that, while remoteness is no protection, the closer you get to heavy industry, the more contaminated you are likely to be. Most humans live very close to the source. In Australia, scientists found more than 4000 chemicals present in sea turtles, seagrass and corals off the Great Barrier Reef, many of which could not be identified. These were increasingly linked to developmental abnormalties in young turtles.[39]

Typical contaminants of marine fish found during scientific studies include lead, cadmium, mercury, arsenic, dioxins, PCBs, polycyclic hydrocarbons, organochlorine pesticides, flame retardants, phthalates, melamine, plastics, marine biotoxins, histamines and radionuclides;[40] most of these substances originate with human industrial activity. While studies typically show that contamination is mostly below the level of concern, researchers express particular caution over the load of heavy metals and organochlorine pesticides carried by long-lived top-predator fish, such as swordfish, tuna, sharks and rays, especially for communities where there is high seafood consumption, as these substances accumulate in both fish and people.

Furthermore, even where fish contaminant levels may be low in toxicity, there is growing concern about contaminants that act as hormone disruptors even in tiny doses, meaning that they can potentially affect the health and development of both the fish and the humans who eat them – especially of the brain, central nervous system, developmental and reproductive systems, at levels far below the toxicity limits set by regulators.

Even the mud on the sea floor is becoming poisonous, with disturbing consequences for marine life – and for the people who eat seafood. Ocean seafood body SeaWeb says: 'Toxic contaminants lead to a severe reduction in the diversity of bottom-dwelling organisms that live in affected estuaries or coastal regions. And adverse effects can spread, via the food chain, to fish, birds, and mammals that feed on contaminated sea life.'[41]

The Norwegian Government adds:

> *Pollutants in sediments can spread to the surroundings. They may spread from the sediment to water, re-suspend when sediments are disturbed, or be absorbed by benthic organisms (bioaccumulation). Because of these mechanisms contaminated sediments may continue to release hazardous chemicals to the surroundings long after the land-based sources of the pollution have been eliminated. As a consequence, contaminated sediments can have serious effects on living organisms and ecosystems.*[42]

Heavy metals, POPs, other toxins and hormone disruptors may kill and interfere with the growth and breeding of sea life. Because many contaminants accumulate in marine food webs, they can also be passed up the chain to humans and so affect our health – as the tragedy of Minamata proved.

People naturally assume that the world's oceans are so vast that even poisonous chemicals, when emitted into the sea, become so dilute as to lose any effective toxicity. For a long time, this false presumption led to the practice of ocean dumping of unwanted materials, such as the disposal of chemical and biological weapons after World War 2, radioactive, industrial and mining wastes, and unwanted pesticides. This resulted in some ocean areas becoming so fouled they had to be treated in the same way as contaminated industrial sites on land. In 2009, for example, the US Environmental Protection Authority announced plans to clean up 17 square miles of ocean off the Palos Verdes Peninsula, California, containing the world's largest dumping ground for unwanted pesticides and PCBs, which were subsequently linked to local disappearances of wildlife.[43]

Mounting scientific evidence throughout the twenty-first century is providing proof that ocean-dumped chemicals are far from safe – and far from gone for good, whatever industry may at times assert. The process known as biomagnification means that toxic substances, originally released in low quantities, can concentrate up the marine food chain until eventually they arrive at human consumers as a poisonous dose. Scientists have shown sharp increases in poisons such as flame retardants in the flesh of Baltic salmon[44] and in deep-sea fish from the Bay of Biscay,[45] methylmercury in fish and marine organisms off China,[46] Alaska,[47] the Florida Everglades[48] and the Canadian Arctic,[49] cadmium in sea creatures off Patagonia and Antarctica,[50] heavy metals off Tanzania[51]... the list goes on and on.

Even in the dwindling fastnesses of the Amazonian rainforest, the claws of contamination are reaching out. Extensive pollution by methylmercury from thousands of small gold workings in the region has been recorded by scientists, affecting both the fish in the rivers and the people who rely on them for food, posing a threat to their health – in a repeat of the tragedy of Minamata.[52]

The presence of these pollutants in deep oceans and distant forests all around the Earth demonstrates that no place on the Planet is now so remote as to be safe from the rising man-made chemical tide, the Anthropogenic Chemical Circulation.

These, and many thousands of similar scientific and official reports, form pixels in a terrifying picture of wildlife both on land and at sea increasingly and extensively afflicted with man-made toxins. They lend growing weight to a view than many otherwise inexplicable population crashes and disappearances – such as the worldwide declines in frogs[53] and honeybees[54] – may be attributable, at least in part, to man-made chemical poisoning. Besides its direct toxicity, our pollution is thought by wildlife scientists to be having a more subtle impact on affected species by stressing their health and fitness to an extent that leaves them more susceptible to diseases, predators or environmental pressures – and by causing genetic and reproductive damage, impairing the chances of survival for many species.

Plastic Surgery

The universal pollution of the Planet has mounted, year by year, almost unnoticed because it is invisible, and humans prefer to trust their eyes over their other senses (including, apparently, their common sense). One form of pollution that is highly visible, however, is plastic.

By 2020 the global petrochemical industry was pumping out around 400 million tonnes of new plastics a year, said the Worldwide Fund for Nature in a trailblazing study.[55]

> *Plastic is not inherently bad; it is a man-made invention that has generated significant benefits for society. Unfortunately, the way industries and governments have managed plastic, and the way society has converted it into a disposable and single-use convenience, has transformed this innovation into a Planetary environmental disaster,*

the WWF bluntly states.

Plastic is valuable for the very reason that, compared with many other materials such as timber, paper or natural fibres, it is highly resistant to rotting and breaking down. Consequently, when it becomes a pollutant, it stays with us for a long, long time – and is rapidly accumulating throughout the Earth system, the oceans and fresh waters especially. Plastic pollution increases at a rate of around 9 per cent a year, driven by an industry that earns over $600 billion annually from it. A study by Roland Geyer and colleagues estimated that, from its discovery in 1907, humanity had produced 8.3 billion tonnes of plastic in total by 2017, less than a tenth of which had been recycled.[56]

Of this vast outpouring, around 8 million tonnes (4–12 mt) is estimated to enter the ocean every year, mostly flushed down the polluted rivers that flow through crowded, dirty cities where thoughtless citizens dump their trash in gutters and rivers. This plays havoc with marine life, as the plastic disintegrates into ever-tinier fragments, eventually becoming a part of the very marine food webs that also sustain human life. European researchers counted up to 1.9 million microscopic plastic pieces *per square metre* on the Mediterranean sea floor off Italy.[57]

A UN conference on plastic pollution estimated that the number of plastic particles in the world's oceans now outnumbers the stars in our galaxy by 500 times, or more. They form a series of five Great Garbage Patches in the main ocean gyres. The worst is the North Pacific Garbage Patch, a gyre measuring from 15 to 20 million square kilometres, estimated to harbour over 3 million tonnes of plastic.[58]

Together these whorls of concentrated plastic waste slay around a million seabirds, 100,000 sea mammals, turtles and countless fish every year when they eat the plastic by mistake. Indeed, Australian scientists found 90 per cent of individual seabirds they examined had eaten plastic and that this 'will reach 99% of all species by 2050'.[59] This is thought to explain, in part, the catastrophic decline in world seabird numbers, with 70 per cent vanishing since 1970.[60] While many people assume ocean plastic consists mainly of large items such as bags and bottles, a growing share of the pollution comes from people's clothing, made from oil-based synthetic fibres. These now make up two-thirds of the world's textiles – and they shed microscopic needles of plastic into the environment every time they are washed: Swedish scientists found that a typical fleece jacket, for instance, releases 7360 invisible microfibre fragments with each wash. Eventually, these flush into the sea.[61]

'Plastic doesn't break down like other organic waste but turns into microplastics which is then easily mistaken by fish and invertebrates as plankton. The consumption of plastics by fish, caught for food for humans, allows plastics to travel up our food chain. Health consequences include ingesting the carcinogenic pollutants that attach to microplastics,' explains the ocean conservation group *Seasave.org*.[62] Medical studies have shown that plastic nanoparticles, if consumed by humans in seafood, can move to any organ in the body – the brain included[63] – causing physical and toxicological damage.[64] Plastic fragments are often poisonous in their own right because of oil-based chemicals used in their manufacture, but when loose in the environment they are also capable of absorbing highly toxic chemicals such as PCBs and polycyclic aromatic hydrocarbons (PAHs)[65] – so acting as a new delivery system, via wildlife, into humans.[66] A study by Lorena Rios and colleagues from the University of the Pacific, California, showed plastic fragments can absorb other contaminants such as PCBs, DDT, PAHs and other POPs, and redistribute these many types of chemicals around the world's oceans.[67]

Driven by ruthless marketplace economics, the danger to both human life and life on Earth generally from plastics is expected to grow significantly towards the mid-twenty-first century. The global switch to renewable energy, the replacement of fossil-fuelled vehicles with electric ones, and the coronavirus pandemic's impact on global transport are driving oil companies away from loss-making fuel production and into increased petrochemical manufacturing, especially of plastics.[68] Consequently, total world output of plastics, and its climate impact, is expected to triple by the mid-twenty-first century (Figure 2.3).

Plastic waste is not only ugly and injurious to wildlife and an engine of climate change, but also serves as a vehicle for the continued transport of man-made poisons around the Planet, in the Anthropogenic Chemical Circulation. Eventually much of this perilous detritus washes up on coastlines or enters the marine – and hence human – food chain.[69]

Out of Sight, Out of Mind

In 1978, a cluster of cancers and birth defects began to appear among residents living in a small suburb at Love Canal, near Niagara Falls, in upstate New York. According to Eckardt C. Beck, one of the US EPA scientists who investigated it, the

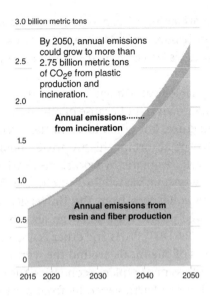

Figure 2.3. Plastics emissions are forecast to triple by the mid-twenty-first century. Source: CIEF.

scene became 'one of the most appalling environmental traged-ies in American history'.[70]

The story began way back in the 1920s, when a disused and dried-up canal was repurposed by the local council as a munici-pal dump. Soon afterwards this was sold to the local Hooker Chemical Corporation and became a disposal ground for all manner of toxic waste. In 1953 the site was earthed over and sold back to the city for $1, who then proceeded to build 100 homes and a school on top of it. In the 1970s, heavy rains, seeping into the soil and flowing subsurface down the old chan-nel, triggered the poison time bomb.

Corroding waste-disposal drums could be seen breaking up through the grounds of backyards. Trees and gardens were turning black and dying. One entire swimming pool had been popped up from its foundation, afloat now on a small sea of chemicals. Puddles of noxious substances were pointed out to me by the residents. Some of these puddles were in their yards, some were in their basements, others yet were on the school grounds. Everywhere the air had a faint, choking smell. Children returned from play with burns on their hands and faces,

Beck recalled, in his vivid first-hand account. After a startling increase in skin rashes, miscarriages and birth defects, President Carter declared a State of Emergency over the site. Beck warned that the ironically named Love Canal was far from an isolated case and there were probably hundreds of similar 'ticking time bombs' all over the USA. State health commissioner David Axelrod presciently described the event as a 'national symbol of a failure to exercise a sense of concern for future generations'.[71]

Such failures persist. The 2000 movie *Erin Brockovich* recounted the story of a crusading legal clerk who exposed the deadly pollution of local groundwater near the town of Hinkley, California, with chromium-6, a cancer-causing contaminant, by a local power company. After years, the episode resulted in a record $333 million legal settlement and the virtual abandonment of the town by its residents. A decade later, a study by the Environmental Working Group (EWG) found chromium-6 in the tap water of thirty-one out of thirty-five US cities, including Washington, Los Angeles, New York and Miami.[72] In 2019 the US EPA's national survey of chromium-6 concentrations in drinking water found the contaminant present in more than three-quarters of water systems it sampled, supplying over 200 million Americans. The EWG noted, 'There is no national standard for chromium-6 in drinking water. The EPA's safety review of the chemical has been stalled by pressure from the industries responsible for chromium-6 contamination.'[73]

Indeed, American environmental laws, designed to protect the lives and health of the US public, were rolled back *en masse* under the regime of President Trump, which favoured the poisoners over their victims: in barely three years, studies by Harvard and Columbia law schools showed that more than 125 environmental laws were watered down or eliminated.[74] This was capped by a decision to increase air pollution by weakening auto emissions standards, at the request of US car makers,[75] in a country where 30,000 people were already dying every year from air pollution. Then, in early 2020, the Trump Administration used the pretext of the coronavirus pandemic to announce a 'temporary policy' of not prosecuting or fining any offending corporation for poisoning Americans.[76] The decision to effectively suspend all environmental laws was taken following a 'call for help' by the American Petroleum Institute and was described by one former EPA official as 'a licence to pollute'.[77] In so doing, the President effectively signed the death warrants for thousands of Americans who will perish as a result of his actions, after Trump himself was long gone.

The failure to uphold environmental laws has itself become pandemic. Today, most of the world's city inhabitants live on top of water so fouled by man-made substances that it is unsafe to drink and, according to UNEP, 'Still today, 80 per cent of global wastewater goes untreated, containing everything from human waste to highly toxic industrial discharges.' This is contributing to a global water crisis, according to *National Geographic* magazine.[78] An estimated 2 billion people drink contaminated water daily.[79]

According to one assessment, the countries with the worst water pollution problems in the world are China, the USA, India, Japan, Germany, Indonesia and Brazil.[80] Cities listed as among the worst for local water contamination include New Delhi, Jakarta, Dhaka, Manila, Beijing, Bogota, Mexico City, Harare, Johannesburg, Tunis, Algiers, Houston, Las Vegas, Jacksonville and Flint. In India – to take one example – the World Economic Forum says, 'As India grows and urbanizes, its water bodies are getting toxic. It's estimated that around 70% of surface water in India is unfit for consumption. Every day, almost 40 million litres of wastewater enters rivers and other water bodies with only a tiny fraction adequately treated.' Scientists estimated that poor water quality was claiming the lives of 400,000 Indians a year and costing the economy $7–9 billion.[81]

Groundwater is the one of the Earth's largest natural resources, accounting for 97 per cent of the Planet's total

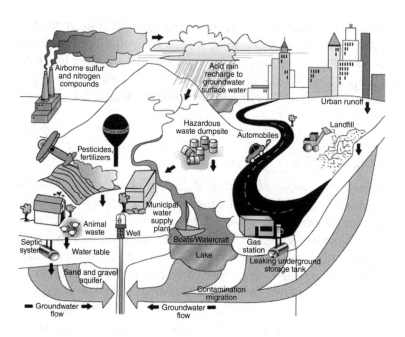

Figure 2.4. Sources of groundwater contamination with man-made chemicals. Source: Groundwater Foundation.

available fresh water. It feeds most of our rivers and lakes, supplies 40 per cent of our drinking water needs and 43 per cent of the irrigation water used to grow food.[82] Groundwater can travel for tens or even hundreds of kilometres underground and is recharged by rainfall on timescales ranging from days to millions of years. Consequently, if it becomes contaminated, the contamination can travel far, in both space and time. The main sources of groundwater contamination are leaky landfills, hazardous waste disposal, illicit industrial discharges and dumping, seepage from old garages and fuel stores, factory sites and gasometers, mining and tailings dams, 'fracking', oil and chemical spills, fire-fighting chemicals, dry cleaning and mechanical solvents, badly managed sewage systems, medications, city runoff and farm chemicals (Figure 2.4).[83] Because it is intimately connected to surface water, groundwater quality inevitably affects drinking water supplies even when these are not drawn directly from wells. Once polluted, groundwater is exceptionally difficult and costly to clean up, unless the pollution is confined to a small area or a single point source, because it often has to be pumped out, cleansed and then reinjected – or else costly underground filtration structures must be built.

Stagnant Seas

Something deeply disturbing is taking place in the world's oceans and estuaries: hundreds of dead zones are forming, areas starved of oxygen and the sea life it supports or awash with algal blooms (Figure 2.5). In recent decades the number of these aquatic black spots has climbed sharply: by 2020 more than 760 such sites had been reported worldwide,[84] up from 50 in the mid-twentieth century. The affected areas lay along the most populous coastlines of Europe, Asia, the Americas and Australia. In all, an international team of researchers calculated, the oceans have lost a total of 77 million tonnes of life-giving oxygen.[85]

Dead zones are hardly new. The first one appeared in the 1850s, when industrialisation killed the River Mersey in the UK. But since that time they have metastasised, steadily and remorselessly invading all the oceans, seas, bays and estuaries most affected by human activity on land.

The cause of dead zones is well understood: their formation is driven by the avalanche of nutrients that humanity dumps in the oceans – from agricultural soil erosion and fertiliser, sewage, leaky landfills, urban stormwater, land clearing, industrial and vehicle emissions. This rich nutrient soup provides the food source for vast blooms of algae, and as these die they sink to the sea floor and decompose, in turn causing blooms of bacteria which then strip the essential oxygen from the water column, resulting in fish kills – their most visible impact. Dead zones are also hastened by global warming which stratifies the water, trapping the stagnant water and preventing it from mixing with the oxygen-rich surface layer.

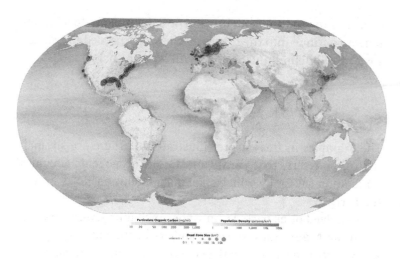

Figure 2.5. World map of ocean dead zones. Source: World Ocean Atlas and R. J. Diaz.

The greatest contributors to these stagnant regions are the 250 million tonnes of fertilisers we use to grow our food, the 75 billion tonnes of eroded topsoil, the sediments unleashed by mining and construction, and fallout from the burning of fossil fuels. These dying ocean regions highlight a deadly paradox: to feed humanity, farm production is expected to increase by 40–50 per cent by the mid-twenty-first century. This means the fertiliser used to grow our food will double too – and so will the destructive impact this has on the oceans. Most affected will be coastal regions, which are already overfished. This means that expanding farming will progressively destroy fishing and fish farming in coastal seas or fresh-water lakes and rivers. A vivid example of this pernicious effect is the dead zone at the mouth of America's Mississippi River, which now extends for more than 15,000 square kilometres (5800 square miles) and is driven by activities such as farming in the river catchment.[86]

Dead zones are one more unsettling manifestation of humanity's Planet-wide chemical impact, arising unintentionally from activities such as farming and mining. They are a clear and present warning that the systems we have employed for centuries to produce food, mineral or industrial products can no longer continue into the future on the scales needed to feed and sustain 10 billion humans. They will have to change.

Six Rivers

The examples in this chapter illustrate that there are six main routes, or rivers, that constitute the Anthropogenic Chemical Circulation (ACC), by which man-made pollution flows around the Earth:

1. Dissolved or as particles in water, including rivers, lakes, groundwater, rain, snowfall and ocean currents.
2. As airborne vapours, gases, microscopic chemical particles or attached to dust particles.
3. In the bodies of living animals and in plants.
4. Via the food chain, which has become contaminated by 1, 2 and 3 and by the intentional use of pesticides, packaging chemicals, food preservatives, dyes and additives in food production.
5. In manufactured goods – traded, transported and used by humans, deliberately or unintentionally, and in their disposal as waste.
6. In humans ourselves, being passed from mother to embryo in foetal blood supply and subsequently in breast milk. From parent to offspring and descendants in damaged genes. And in the drugs we take in.

These pathways illustrate how human chemical pollution flows around the Planet in the ACC. They also highlight the near-impossibility of controlling its spread anywhere except at its source – or very close to it. This underlines the urgency of science properly defining the scale of the problem – as it has for carbon – and its primary sources, in order that they be controlled.

And while the Earth is a large place and pollution is still fairly dilute, pollution levels are poised to increase several-fold by the mid-twenty-first century under the pressure of rising demand for food, housing, energy, industrial goods and services, driven by population growth and demand for higher living standards.

Toxic Blindfold

These examples of the pervasive impacts of chemical pollution – and tens of thousands more like them, documented in scientific reports – mean that human-sourced contamination is now a universal scourge impacting all nations, all societies and the health and wellbeing of every person – no matter where they live – as well as most forms of life on Earth.

In its first Global Chemical Outlook (2013) the UN Environment Programme cautioned:

> *Environmental effects of the chemical intensification of the national economies are ... compounded by the trans-boundary movement of chemicals through the air or water. In some countries this occurs because they lie downriver or downwind from the polluting industries of neighbouring countries. In other countries, the runoff of pesticides and fertilizers from agricultural fields or the use of chemicals in mining in neighbouring countries, may leach into ground water, or run into estuaries shared across national boundaries. Throughout the globe, atmospheric air currents deliver chemical pollutants which originate from sources thousands of kilometres away.*
>
> *Whilst each chemical-intensification factor contributes to a small share of the environmental burden of each country and nation state, when combined, these together can form an increasingly significant and complex overall mix of chemicals not present fifty years ago. As this chemical intensity increases, the prospects for widespread and multifaceted exposures of humans and the environment to chemicals of high and unknown concern also arise.*[87]

In its second report, released in 2019, UNEP director Joyce Msuya said, 'Global trends are rapidly increasing chemical use ... whether this trend becomes a net positive or a net negative for humanity depends on how we manage the chemicals challenge.

What is clear is that we must do much more ... We cannot live without chemicals. Nor can we live with the consequences of their bad management.'[88]

The UNEP report, be it noted, refers here only to the 2.5 billion tonnes of man-made chemicals, not the 120–220 billion tonnes embodied annually in the Anthropogenic Chemical Circulation due to activities such as mining, farming, construction, manufacturing, power generation and development.

To that, the world remains, for the most part, blindfolded.

This combined chemical assault on Planet Earth is an issue for which no nation, industry, corporation or society presently accepts full responsibility, even for its own share of the mess.

It is an issue to which most individuals are blind – including those who release chemicals – even when it comes to protecting their own families. It is as if humanity has taken a profound collective decision that it is more important for us to consume than to protect our health or that of our children. This is a decision that carries stark implications for our own survival and wellbeing as a civilisation and as a species.

There is no future on a poisoned Planet.

3 ARE YOU A CONTAMINATED SITE?

'It is ironic to think that man might determine his own future by something so seemingly trivial as the choice of an insect spray.'

Rachel Carson, *Silent Spring*, 1962

In a run-down hospital in Bihar, India, five-year-old Rashmi Kumari was fighting for her life. She was the only child of her family to remain alive; twenty-two of her school friends had already perished. Journalist Rajesh Roy reported in *The Wall Street Journal*:

> Inside an ambulance, an Indian man cradles his dead daughter, who had fallen ill after eating a school lunch. Initial investigations suggest that organophosphorus, commonly used in farm pesticides, may have been mixed into the rice, beans and potato curry served at an elementary school in Gandaman, a village in the impoverished state of Bihar, according to Amarkant Jha Amar, medical superintendent at the hospital.

> The students became sick and suffered from vomiting, fainting and foaming at the mouth. More than two dozen victims are still being treated. The state's education minister, P.K. Shahi, said the school's cook noticed a strange color and foul smell from the cooking oil when preparing lunch. Children also complained about the food, he said. He said the cook informed the school's headmistress. After tasting the food, the cook also became ill and was hospitalized.[1]

News of the Indian school children's poisoning rang out around the world. Adding to the horror was the fact that the poison was sourced to a government-sponsored school lunch programme intended to help poor families by providing better nutrition for their children. Someone, it seemed, had put the food in old, unwashed pesticide containers.

Just another unfortunate accident? Or the inevitable result of a system that relies on chemicals in all aspects of daily life and has become over-casual about their use? And, looking closer, whose school lunches are truly free from all forms of industrial chemical, toxin or pesticide? Whose children are free of such things? Have we seriously looked?

***

Society is accustomed to thinking of pollution in terms of contaminated sites: areas of land too fouled by industrial leftovers to be safely used. Yet mounting scientific evidence from around the world suggests that today each individual is their own contaminated site. From the moment of conception to the moment of death, we all now accumulate a potentially debilitating and sometimes lifelong store of poisonous substances that originate in our consumption and living patterns and the places we inhabit. Many of these substances can now be identified by a range of common, though expensive, tests. This is not a matter for belief, denial or ideology: if you doubt it, get your blood, hair, saliva and urine thoroughly checked.

The world was awakened to the risks of profligate use of toxic chemicals by American biologist and writer Rachel Carson, when she published her celebrated book, *Silent Spring*, in 1962,[2] raising a concern that helped ignite the global environmental movement. Carson's focus was on the excessive use of long-lived pesticides such as DDT in the food chain and environment at the time – yet pesticide use has since increased more than tenfold to around 5 million tonnes globally today. Once employed chiefly in developed-country agriculture, the use of pesticides has spread around the world, driven chiefly by supermarket and food company demand for huge volumes of cheap, uniform, blemish-free produce – with the result that few, if any, consumers of the modern industrial diet would find themselves uncontaminated if tested.

Yet pesticides, despite the high public focus and media scrutiny they receive, are in fact only the mere tip of the chemical iceberg, most of which remains out of sight and out of mind.

So immense has the task now become of measuring – in either the environment or in humans – the true extent of contamination from both legally manufactured and unintentionally released substances that most national regulators and governments do not even attempt it. Instead they confine themselves mainly to limited investigations of a few chemicals in cases of immediate concern.

This means that, at best, they are examining only a handful of pieces from a vast jigsaw puzzle, revealing only dissociated fragments of the image it portrays. We do not know the big picture or anything approaching it.

Polluted People

A chilling glimpse of the big picture comes from the USA, among the heaviest chemical users on the Planet. For more than two decades its Centers for Disease Control (CDC) has run a national survey of chemical pollution in the blood, serum and urine of up to 2500 Americans every year. Its aims include determining which noxious chemicals are entering US citizens and at what concentrations, how many people are contaminated above the danger level, and which groups are most at risk.[3] Recent tests cover 337 substances of known or suspected toxicity,[4] out of the 86,000 chemicals officially sanctioned for use in the USA.

Basically, the survey reveals that Americans are a walking cocktail of contaminants: 'In the majority of individuals tested, acrylamides, cotinine, trihalomethanes, bisphenol A, phthalates, chlorinated pesticides, triclosan, organophosphate pesticides, pyrethroids, heavy metals, aromatic hydrocarbons, polybrominated diphenyl ethers, benzophenone from sunblock, perfluorocarbons from non-stick coatings, and a host of polychlorinated biphenyls and solvents were found.'[5] And those are *just a handful* of the substances that the testing covers. As to the main sources: 'For most chemicals, people are exposed to low levels in foods they eat, water they drink, and the air they breathe. People can also be exposed to chemicals from products and containers they use,' says the CDC.[6]

To take a single example, phthalates, which are substances used to improve the durability of plastics: 'CDC researchers found measurable levels of many phthalate metabolites in the general population. This finding indicates that phthalate exposure is widespread in the U.S. population. Research has found that adult women have higher levels of urinary metabolites than men for those phthalates that are used in soaps, body washes, shampoos, cosmetics, and similar personal care products.'

They also found 'measureable mercury in most of the participants [in the survey]', that Americans had '3 to 10 times higher than levels [of PBDEs] seen in ... European countries', that perchlorate used in rocket fuel was present 'in all participants', that PFOA was found 'in the serum of nearly all the people tested, indicating [it] is widespread in the U.S. population', and acrylamides 'in the blood of 99.9% of the U.S. population'.[7]

What this means at a personal level is described by journalist David Ewen Duncan in a telling account in *National Geographic* magazine:

Last fall I had myself tested for 320 chemicals I might have picked up from food, drink, the air I breathe, and the products that touch my skin — my own secret stash of compounds acquired by merely living. It includes older chemicals that I might have been exposed to decades ago, such as DDT and PCBs; pollutants like lead, mercury, and dioxins; newer pesticides and plastic ingredients; and the near-miraculous compounds that lurk just beneath the surface of modern life, making shampoos fragrant, pans nonstick, and fabrics water-resistant and fire-safe.

The tests cost Duncan $15,000 and told him a whole lot of things he really didn't want to know about himself – that he contained a toxic shower, parts of which dated back to his time in the womb, to his youth – he had lived near heavy industry – and his recent life in touch with an ever-expanding array of suspect substances.[8] Disturbingly, no-one could tell him what it all meant.

The CDC cautions that the presence of a certain chemical in blood or urine need not mean the health of the individual is at risk – and that the health effects of many of the chemicals in its survey are not yet clear. However, it also admits that the threshold levels for adverse health effects of many chemicals are simply not known. And it is silent about the possible health effects of complex mixtures of chemicals and their breakdown or reaction products.

America is the second largest user of chemicals in the world, after China. The next four big users are Germany, Japan, India and Brazil.[9] Without comprehensive regular surveys like the CDC's, it is impossible to say how polluted are their citizens – and this surveys less than 0.5 per cent of all chemicals in use. But there is little doubt that the picture is similar in virtually every country on Earth where modern agriculture, household products, cosmetics, vehicles and pesticides are commonly used.

The longer we live, the worse the problem gets. Emeritus Professor Michael Depledge of the University of Exeter Medical School told the UK House of Commons Environmental Audit Committee: 'As we are now living longer, we are accumulating levels in our bodies that are much higher than ever before, so there is a much larger number of people with higher levels of these chemical mixtures than ever before and we do not know what the implications are of it.'[10]

Acknowledging the vastness of the problem, the World Health Organization (WHO) commented:

> The number of existing chemicals and their compounds is very large, and for many of them the health risks are not known. Chemicals can be the result of anthropogenic sources [caused or produced by humans] or occur in nature. Hazardous chemicals can reach our body through different routes (e.g. food, air, water) and cause a variety of health effects. Due to the many ways in which chemicals are used and released, the many exposure routes involved, and the different mixtures of chemicals present, the public health relevance of chemicals can be extremely difficult to assess.

Exposure to and accumulation of chemicals continues in most people throughout their lives. To take but one example, the WHO considers that 'all people [author's emphasis] have background exposure and a certain level of dioxins in the body, leading to the so-called body burden.' While this may not cause serious illness in most people, nevertheless the risks associated with dioxin exposure include skin diseases, altered liver function and cancers. The effects are far more dangerous for the unborn child. Dioxins are found throughout the world in the environment and they accumulate in the food chain, mainly in the fatty tissue of animals.[11]

Of dioxin in particular the WHO says:

More than 90% of human exposure is through food, mainly meat and dairy products, fish and shellfish. Dioxins are highly toxic and can cause reproductive and developmental problems, damage the immune system, interfere with hormones and also cause cancer. Due to the omnipresence of dioxins, all people have background exposure, which is not expected to affect human health. However, due to the highly toxic potential of this class of compounds, efforts need to be undertaken to reduce current background exposure.[12]

Concerned at the gaps in our knowledge, the Environmental Working Group (EWG), a not-for-profit American NGO, ran eleven separate studies of its own, with the aim of finding out how polluted Americans are. It established they were affected, literally, cradle to grave. 'These projects, employing leading biomonitoring labs around the world, have together identified up to 493 chemicals, pollutants and pesticides in people, from newborns to grandparents,' it said.[13]

BPA Babies

In a ground-breaking report the EWG also revealed that Americans enter the world already burdened with a cargo of toxic industrial chemicals: 'A two-year study involving five independent research laboratories in the United States, Canada and the Netherlands has found up to 232 toxic chemicals in the umbilical cord blood of 10 babies from racial and ethnic minority groups. The findings constitute hard evidence that each child was exposed to a host of dangerous substances while still in its mother's womb,' the Group said.

Nine out of the ten baby samples contained bisphenol A (BPA), an industrial petrochemical produced by the millions-of-tonnes annually in the production of polycarbonate plastics and epoxy resins: 'BPA has been implicated in a lengthening list of serious chronic disorders, including cancer, cognitive and behavioral impairments, endocrine system disruption, reproductive and cardiovascular system abnormalities, diabetes, asthma and obesity.' Sources of BPA include plastic containers, canned food, toiletries, female hygiene products, thermal printer receipts, CDs and DVDs, household electronics, food cartons, timber particle board and furniture made with epoxy glues. Bizarrely, most of the babies' blood also contained traces of chemicals used in explosives and rocket fuel, which the EWG attributed to poorly maintained defence facilities allowing such chemicals to leak into groundwater, which supplies much of America's drinking water. Additionally, industrial chemicals used to make cosmetics, detergents, soap and other scented household items were found in seven out of ten samples.[14]

'Each time we look for the latest chemical of concern in infant cord blood, we find it,' said Dr Anila Jacob, EWG senior scientist and co-author of the report. 'This time we discovered BPA, among other dangerous substances, in almost every infant's cord blood we tested.'

The EWG studies shone a harsh spotlight on concerns that doctors and medical scientists had harboured for decades. The first national studies designed to find out how poisoned are modern human babies were run by Norway and Denmark in the late 1990s. In 1997, leaders of the G8 group of countries meeting in Miami declared 'We acknowledge that, throughout the world, children face significant threats to health from an array of environmental hazards', specifying 'chemical contaminants in drinking water, air pollution that exacerbates illness and death from respiratory problems, polluted waters, toxic substances, pesticides, and ultra-violet radiation'. And mentioning 'emerging threats to children's health from endocrine disrupting chemicals'.[15] The USA, UK, France and South Korea started big studies – though the USA and UK soon abandoned theirs. Japan launched a long-term national birth cohort study involving 100,000 mothers and their children, the *Japan Environment and Children's Study*, in 2010, but does not expect to announce results before 2027. A global co-operation group was formed to help everyone adopt the same approach so results can be compared globally.[16] And, round the world, study after study reveals disturbing links between chemical poisoning and diseases in children.[17]

The peril that babies and young children now face was summarised by William Suk of the US National Institutes of Health and colleagues in a study that found 'Exposures to environmental pollutants during windows of developmental vulnerability in early life can cause disease and death in infancy and childhood as well as chronic, non-communicable diseases that may manifest at any point across the life span.' In low-income countries it was a 'major threat', both to children and to entire economies, because of the burden of ill-health it imposes.[18] They described it as an 'under recognized threat', which is science-talk for 'you are not paying attention'.

Yet, despite all the good intentions, the warnings and the high-minded statements, the poisoning of humanity's children continues to grow – with the increase in synthetic chemical production, use and release.

Today's newborn infant has barely drawn its first few breaths before it receives a mouthful of pesticide and flame retardants. A team at UC Berkeley found pesticides and PCBs in all samples of mothers' breast milk it collected in the San Francisco Bay area and in the Salinas Valley: 'These results suggest that neonates [newborn babies] and young children may be exposed to persistent and non-persistent pesticides and PCBs via breast milk,' the researchers concluded. The most disturbing feature of this study

was that most of the pesticides found were the modern, sup-
posedly 'non-persistent' pesticide types – in other words, those
introduced specifically to replace banned 'persistent' organochlorine chemicals – and whose health effects remain 'largely
unknown' according to the CDC.[19]

Similar findings are echoed in *hundreds* of scientific studies
around the world: 'human milk contamination by toxic chemicals such as heavy metals, dioxins and organohalogen compounds, however, is widespread and is the consequence of
decades of inadequately controlled pollution'.[20] A Chinese study
reported traces of organochlorine pesticides such as DDT and
lindane were found at high levels in mother's milk in
Guangzhou.[21] A study in the Punjab region of India also found
disturbing levels of a wide range of pesticides and agricultural
chemicals.[22] In Europe, researchers reported dioxins, PCBs and
other POPs in the breast milk of nursing mothers from fifteen
different countries in four separate surveys run between
1987 and 2007.[23] Fluorine-based chemicals, used widely in fire-
fighting, are among common contaminants of breast milk in
Europe, America, Asia and Australia, and are often poorly regulated.[24] A group called Moms Across America commissioned
research which found the world's most widely used herbicide,
glyphosate, in 30 per cent of samples of breast milk analysed. It
was present at levels up to 1000 times higher than permitted in
drinking water in the EC, leading for calls for a ban on its use in
food crops.[25]

'Mums in the UK have some of the world's highest concentrations of flame retardants in their breast milk, some of which
have now been banned,' Mary Creagh, chair of the UK House of
Commons Environmental Audit Committee told the British
Parliament and media in 2019. 'Chemical flame retardants are
still being widely used in our furnishings from children's mattresses to sofas. Meanwhile the Government is sitting on its
hands instead of changing regulations to ensure that the most
toxic chemicals are taken out of use.'[26]

A review carried out for the Stockholm Convention on
Persistent Organic Pollutants (POPs) concluded that these

pesticides and industrial chemicals are 'distributed to women in all parts of the world and are thus delivered to the nursing child', adding that 'the nursing child is targeted by a vast number of undesirable pollutants'. Many of these substances are highly stable, meaning they don't break down easily and can accumulate in humans, especially nursing mothers, and then be passed to their babies in ever-increasing doses. The Stockholm study found PCBs and dioxins were higher in nursing mothers in Europe and North America, whereas pesticides were higher in Africa and Asia, and PBDEs were higher in the USA.[27]

While levels of contaminants in mother's milk vary from country to country, a study co-ordinated by the UN Environment Programme in fifty-two countries found 'A risk–benefit assessment indicates that human milk levels of PCDDs, PCDFs and PCBs are still significantly above those considered toxicologically safe.'[28] It concluded that, where breastfeeding is concerned, the benefits still outweigh the risks – but there is 'a strong argument ... to reduce human exposure further to dioxin-like compounds'.

Companies that produce pesticides, plastics and other contaminants generally insist that *their* products are harmless in small doses, while turning a conveniently blind eye to the overall toxin load delivered to newborns from a multiplicity of sources. However, as countless scientific studies have confirmed, such claims are less than truthful. As nutritionist M. Nathanial Mead bluntly put it: 'Breastfeeding infants are ... the final target of POPs.'[29]

A major source of the contamination for newborns is 'air toxics', many of which are emitted by the home and its furnishings, then inhaled by the mother and passed to her baby via her bloodstream or milk. These substances are emitted from synthetic fabrics, carpets, fibreboard, plastic products, glues and solvents, spray cans, printed materials, paints, varnishes, cleaning products, disinfectants, cosmetics, degreasing products, hobby products, insect sprays, tobacco smoke, fuels, vehicle interiors, home heaters and fireplaces, to name just some sources. Indoor emissions can occur at much higher levels than in polluted city air: families who use an indoor air quality monitor are often shocked at the rise in air pollution that occurs in a room the moment the window is closed. Since modern urban citizens spend over 90 per cent of their time indoors, exposure to indoor air toxics is now regarded as a serious and lifelong potential source of disease, as these toxics may be inhaled with virtually every breath. Mothers and their unborn or breastfeeding babies are especially exposed in the modern home.

Other potential sources of contamination of the human foetus and newborn are the now-ubiquitous cosmetics and perfumed soaps. The EWG found phthalates, triclosan, parabens and musks used in common 'personal care' products in the umbilical cord blood of newborn babies, showing clearly that they had been contaminated while still in the womb, as well as subsequently in the air they breathed and the milk they drank.

Together, these many studies make it clear that virtually every infant on the Planet is now being poisoned from its first day of life by unwanted industrial substances in its mother's milk. They show that breast milk, that uniquely precious and revered fluid, is contaminated worldwide and that homes, where people think they are safe, are a major source of pollutants. And while bans have been effective in reducing a few contaminants, such as DDT, the overall problem is becoming worse owing to rapidly increasing global chemical use throughout society and the substitution of new, untested, chemicals for the ones being withdrawn.

Baneful Beauty

One of the first things that the modern citizen does upon waking each morning is try to get cancer.

When people shower, they bathe themselves in chemicals from head to foot, many of which are suspected or known to be toxic, carcinogenic or allergenic. They then go on to cover themselves in a positive chemistry set. Today's shampoos, soaps, bath gels, cosmetics, lotions, hair sprays, deodorants and perfumes contain a witches' brew of contaminants – most of them derived from petroleum – that add greatly to the ongoing total exposure of both the user and everyone around them, including babies and children.

'Over a lifetime you will potentially swallow a kilogram of lipstick,' author Dawn Mellowship cautioned her readers, noting the average women used twelve personal care products a day, every year she absorbs about 2 kilograms of cosmetics into her body and that, worldwide, 'use of cosmetics is out of control'. 'I freely admit that, in days gone by, I was a toiletry and cosmetic addict . . . that changed when I developed a chronic illness that caused me to re-evaluate my life and rethink my priorities.'[30] Yet, today, cosmetics are a growing industry worth over $700 million a year.[31] Disingenuously, this is referred to as the 'personal care' industry. On the basis of the emerging science, 'personal harm' is closer to the mark.

The University of Toronto, which conducts one of the world's leading research programmes into cosmetics, says:

> Exposure to perfumes and other scented products can trigger serious health reactions in individuals with asthma, allergies, migraines, or chemical sensitivities. Fragrances are found in a wide range of products. Common scented products include perfume, cologne, aftershave, deodorant, soap, shampoo, hairspray, bodyspray, makeup and powders. Examples of other products with added scents include air fresheners, fabric softeners, laundry detergents, cleaners, carpet deodorizers, facial tissues, and candles.

> We generally think that it is a personal choice to use fragrances; however, fragrance chemicals are by their very nature shared. The chemicals vapourize into the air and are easily inhaled by those around us. Today's scented products are made up of a complex mixture of chemicals which can contribute to indoor air quality problems and cause health problems.

> Some of these fragrance chemicals are known to be skin sensitizers. Some are also respiratory tract irritants and can trigger asthma and breathing difficulties. Asthmatics commonly cite fragrances as

initiating or exacerbating their asthma. Fragrances are also impli-
cated in vascular changes that can trigger migraines in susceptible
individuals. Individuals with chemical sensitivities can experience
symptoms at very low levels in the air, far below those known to cause
harmful effects in the general population. Susceptible individuals can
experience a variety of symptoms, including headache, sore throat,
runny nose, sinus congestion, wheezing, shortness of breath, dizziness,
anxiety, anger, nausea, fatigue, mental confusion and an inability to
concentrate. Although the mechanisms by which fragrance chemicals
act to produce symptoms are not yet understood, the impact on all
those affected can be quite severe, resulting in great difficulty in work
and study activities.[32]

Concerned at the lack of official scrutiny or regulation of this
popular but often poisonous form of self-adornment, the EWG
publishes an extensive list of 9000 ingredients making up 152,000
unique chemicals commonly found in 85,000 'personal care'
products, to better inform consumers about health risks when
making their choices. Noting 'the US Government doesn't review
the safety of products before they are sold', the EWG tries to help
the baffled consumer navigate the chemical thickets of cosmetics
by providing a 'safety scorecard', based on what the manufacturer
says is in their products and what government databases say
about the ingredient's toxicity. However, this relies on the manufacturer listing product ingredients truthfully and accurately on
their label.[33] Even then, says EWG, '595 cosmetics manufacturers
have reported using 88 chemicals that have been linked to cancer,
birth defects or reproductive harm in more than 73,000 products.' The Group called for a ban on their use in cosmetics.[34]

Globally the situation is riskier still: many countries that
produce cosmetics do so without any regulation or official scrutiny. So ubiquitous have these products become that every international airport now presents a gauntlet of perfume and
unguent sales outlets that must be run before the passenger
can reach their flight. Like the alcohol and tobacco peddled
alongside them, these luxuries may inflict serious harm – but,
in contrast, warnings about their use are invisible.

'While the chemicals in cosmetics make us look, feel, and smell better, research strongly suggests that at certain exposure levels, some of these chemicals may contribute to the development of cancer in people,' states the USA-based non-profit Breastcancer.org, founded by Dr Marisa Weiss, which reaches over 200 million women worldwide. It adds, 'But because personal care products contain a diverse combination of chemicals, it's nearly impossible to show a definite cause and effect for any specific chemical on its own.' Another online group, the Campaign for Safe Cosmetics developed by organisations concerned with breast cancer, also publishes useful consumer advice on toxic compounds in cosmetics to avoid.[35]

Like the 7000 chemicals in tobacco smoke, it is nearly impossible to say which cosmetic ingredients cause harm to the health and wellbeing of their users or what effect their mixtures and combinations with other pollutants in air, food and water may have. Also, since the 'cosmetics craze' is a relatively recent global phenomenon, the longer-term warning signs are not as apparent or familiar to consumers as they are for tobacco or alcohol, which have been in use for centuries. Dr Shuai Xu of Northwestern University and his team analysed over 5000 cases of reported cosmetic harm reported to the US Food and Drug Administration (FDA) between 2004 and 2016. 'The data suggest that consumers attribute a significant proportion of serious health outcomes to cosmetics,' they said, noting that US cosmetics manufacturers were under no legal obligation to report cases of harm, citing one case where a hairspray maker had received 21,000 complaints, none of which they reported. They concluded there was evidence of a rise in harmful events associated with hair care products, skin care products and tattoos especially, adding, 'There have been multiple public health controversies surrounding cosmetics involving lip balms, lipsticks and eyelash makeups adulterated with prostaglandins.' The most serious health impacts were reported for baby care, personal care, hair care and hair colouring products.[36]

There is a shattering contrast between US and European scrutiny of chemicals in cosmetics, making health and safety all the harder for the consumer to navigate. The US FDA has banned eleven cosmetic substances from use.[37] The European Commission has banned 1623.[38] Most countries that police cosmetics follow one or the other, making this a matter for debate worldwide whether chemical regulation exists for the benefit of consumers – or for manufacturers.

However, driven by informed consumers, some international cosmetics firms are responding by removing many of the dangerous substances from their products – but this seldom applies to locally made cosmetics and so-called 'natural' brands. Australian consumer group CHOICE says, 'On the upside, we found many of the products from major international brands no longer contain any of the more dubious ingredients. As a result, the once-common dibutyl phthalate, toluene, butylated hydroxyanisole (BHA) and petroleum distillates have all but disappeared from big brand nail polishes, lipsticks and mascaras. However, beyond these international brands, the findings were less reassuring.' In cheap retailers they found 'examples of cosmetics made in Australia, Asia and the Middle East that contain chemicals banned or restricted elsewhere'.[39]

Owing to a lack of generally accepted international rules for cosmetic safety, however, it is unclear whether the new chemicals that have been substituted for known toxins are safe, as there is no requirement for them to be tested. Frequently, as illustrated by HCFCs and pesticides, the substitute turns out to have as many risks as the original substance, though they may be different risks. Furthermore, there are very few studies that address questions such as the overall body burden delivered by cosmetic chemicals or its effect on health, including on the hormonal, gender and reproductive systems of women and their babies.

However, given the pleasure that these products evoke, it may be difficult for people who enjoy using perfume and cosmetics to appreciate they are in fact risking self-poisoning or exposing their babies, family, friends and colleagues to toxic risk in a manner scarcely different to tobacco smokers. The toxicity that, on average, people widely recognise in the cases of tobacco, alcohol, drugs and food is, in the case of the deceptively named 'personal care' products, blithely ignored.

Yet cosmetics use underlines a vital point about the nature of global contamination: this contamination arises not only from the unscrupulous or indifferent actions of large corporations and the seemingly benign neglect of governments – but also from the combined wishes, preferences and choices of billions of consumers who are prepared to risk trading their health, and that of others – including their children – for the immediate gratification that such luxuries afford.

Breast cancer seems a high price to pay for great hair.

Food Fears

North Carolina mother Lisa Leake was watching food author Michael Pollan being interviewed about his latest book on *The Oprah Winfrey Show* when, she freely admits, she had an epiphany. 'I was very intrigued and went on to read his book *In Defence of Food*. It was a huge wakeup call to me. I couldn't sleep at night when I was uncovering all this information ... I was appalled because I thought I was feeding my family healthy food – but the foods I thought were healthy turned out to be highly processed.'[40]

Like many innocent consumers, Lisa had assumed there were actual strawberries in strawberry syrup and lemons in lemonade, only to discover to her horror the flavourings were in fact made from synthetic chemicals. It was the start of an interest that became a fascination and then a global crusade, in the form of Lisa's *100 Days of Real Food* website and blog,[41] which has attracted millions of fans worldwide and provides advice about what 'real food' is and where to get it. She converted her family's diet to one of whole, fresh foods, noting immediate improvements in health. Her infant daughter who had suffered constipation and asthma was freed of both ailments. 'We all had a decrease in sicknesses over the next year,' she said. Her own blood cholesterol reading plummeted spectacularly in the healthy direction. To show that it is indeed possible to eliminate many of the deleterious substances from our modern diet, she designed her 100-day Real Food Challenge, where families commit to eating mainly whole grains, fresh fruit and vegetables, locally raised meats and other wholesome foods that have not been processed or had things added to them. She adopted Michael Pollan's advice: read the label first and never eat any food containing a substance you can't pronounce or you don't trust. She advises her fans to patronise the local farmer's market and a local bakery that uses wholegrain ingredients, instead of the supermarket. What began as one concerned mother's obsession has become a movement that is rapidly spreading around the world.

Lisa is just one among millions of consumers globally who are wising up – and shunning the modern industrial food supply, fearing it is a source of lifelong personal contamination. Due to the extensive (and growing) use of chemicals to grow, preserve, protect, process, sweeten, dye and flavour at least half of the world's food, the modern diet is increasingly recognised by scientists and health experts as risky. The World Health Organization explains, 'Chemicals can end up in food either intentionally added for a technological purpose (e.g. food additives), or through environmental pollution of the air, water and soil. Chemicals in food are a worldwide health concern and are a leading cause of trade obstacles.'[42]

Over 1000 different chemical pesticides are used in agriculture worldwide, with the aim of controlling pests or weeds that can undermine the food supply.[43] Regardless of claims made about their safety, they are nearly all designed to kill something, and are therefore poisonous to many life forms, including people. Many of them are long-lasting, and can 'bioaccumulate' – that is, build up their dose in living organisms which consume them, including people. Around 5 million tonnes of pesticides are used to grow food crops and livestock globally every year (Figure 3.1), and the volume is rising steadily, worldwide.[44] Scientific studies show that two thirds of the world's farmlands are now contaminated by pesticides.[45] These chemicals sometimes leave trace residues in foods, both fresh and processed, which can be swallowed by consumers. In the low concentrations found in a single food

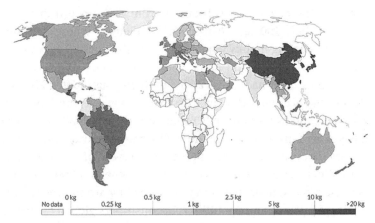

Source: UN Food and Agricultural Organization (FAO)

Figure 3.1. World pesticide use per hectare of cropland, 2017. Source: UN FAO/Our World in Data.

item – as their manufacturers often claim – individual substances are usually deemed harmless. In higher concentrations, however, many can be harmful to human health, and have been scientifically tied to cancers, brain, nerve, genetic and hormonal disorders, kidney and liver damage, asthma and allergies. Babies exposed to pesticides while still in the womb may later suffer childhood cancers, brain/nerve and behavioural problems.[46] The effects of regular, low doses of many different pesticides in food over years or a lifetime are unknown, as are the health effects of small doses of pesticide in mixtures.

Pesticides are among the leading causes of death by self-poisoning, especially in developing countries. Farm workers are most at risk, while consumers are exposed at far lower levels. However, 80 per cent of the increase in world food output needed to feed the global population in 2050 will require the use of more pesticides, the WHO foreshadows.

The pesticide issue is vividly illustrated in a long-running battle between river conservationists and a big supermarket salad supplier in Hampshire, UK. To cleanse fresh vegetables from pesticide, supplier Bakkavor washed them in water from the local River Alre, then discharged the water back into the river, a tributary of the Itchen, a river famed for its insect and fish life, which had declined in recent years. Pesticide levels in the released water were up to 400 times the maximum permitted levels, the local fish conservation body said – a claim disputed by the supplier.[47] There are over a quarter of a million supermarkets worldwide servicing 5 billion consumers: that each of them is helping to poison a river somewhere merely by washing salad in its water gives some notion of the toxic machinery behind your 'fresh' vegetables, and its wider consequences for the environment.

Besides pesticides, some 3000 chemical food ingredients are permitted in the USA by the Food and Drug Administration; this is a fair indicator of what goes on worldwide, since many countries uncritically emulate the US industrial food model, and adopt permissive US regulations. Europe, taking a rather more cautious view, lists 400 permitted food additives and specifies how they can be used.[48] These additives are used to preserve or enhance food safety, freshness, taste, texture or appearance. The WHO says they should all be checked for potential harmful effects on human health before use. Only those deemed 'safe' can, in theory, be used in global trade.

To highlight one example, much industrial food today contains dyes and bright colourings, many of which have been scientifically linked to, or are suspected of causing cancer and other health issues, particularly in children. University of California LA molecular toxicologist Sarah Kobylewski observes: 'Dyes are complex organic chemicals that were originally derived from coal tar, but now from petroleum. Companies like using them because they are cheaper, more stable, and brighter than most natural colorings.' In her report *Food Dyes: A Rainbow of Risks*, she pointed out that Americans alone may eat around 3000 tonnes of these dyes a year, most of which have not been properly health tested – except, in some cases, by the people who made them. Consumption of dye increased sixfold from 10 milligrams per person per day in the 1950s to 60 milligrams in the 2000s, making it likely that the consumption of carcinogens –

which contaminate some common dyes – had also reached
unsafe levels. She pointed out that several US multinational food
firms were selling the same food products in the USA and
Europe, with only the US version containing dyes. 'Most of those
companies said that they don't use dyes in Europe because
government has urged them not to – but that they would continue to use dyes in the United States until they were ordered not
to or consumers demanded such foods,' she noted.[49]

Another category of public concern is the almost universal use
of 'preservatives' to extend the shelf-life of processed and fresh
foods. These are generally chemicals that poison the bacteria
and moulds which cause food to rot, and therefore help save many
lives from food poisoning. However, common chemical preservatives such as sodium nitrate and nitrite, sulphites, sulphur dioxide,
sodium benzoate, parabens, formaldehyde and antioxidant
preservatives, if overconsumed in the modern processed food diet,
may also lead to cancers, heart disease, allergies, digestive, lung,
kidney and other diseases and constitute a further reason for
avoiding or reducing one's intake of 'industrial food'.
Consequently, many researchers are now seeking safer 'natural'
compounds to preserve food.[50] Preservatives furnish an example of
how the chemical industry seeks to mislead the public: on the
industry-sponsored website *ChemicalSafetyFacts.org* the benefits of
chemical preservatives are explained – but none of the risks are
acknowledged.[51]

To these intentional additives, we must also consider several
thousand 'indirect food additives' or food contact chemicals,
which become part of the food during its processing, packaging,
storage or handling. These mostly leach out of plastic and
treated paper containers, cling wraps, cans, plastic bags and
foam containers – and mostly originate with the fossil fuels used
to manufacture the packaging. In 2020 a warning about these
was issued by thirty-three leading scientists in a consensus statement: 'Food contact chemicals transfer from all types of food
contact materials and articles into food and, consequently, are
taken up by humans ... We are concerned that current chemical

risk assessment for food contact chemicals does not sufficiently protect public health.'[52]

Nobody knows for sure how many chemicals have been added to the human food supply. Back in 1983 the US National Institutes for Health estimated the figure at 14,500, of which 12,000 were indirect additives.[53] This number did not include agricultural pesticides or natural contaminants such as aflatoxins. However, the number has only grown with the increase in the range of synthetic chemicals used in agriculture, food processing and packaging over the intervening four decades. Today the number is likely to be well over 16,000.

'Over the last few decades, the number of chemicals added to foods and other products has skyrocketed,' says Harvard Health's Dr Claire McCarthy.

We have created all sorts of plastics that are used in innumerable ways. We add preservatives to foods to keep them fresh. We add chemicals to foods to make them look more appealing. We have made food packaging to keep food fresh. We add chemicals to lotions and beauty products to make them feel, look, and smell nice ... the list goes on and on of the ways we have invented and used chemicals. We did all of it for what seemed like good reasons at the time, but we are learning that many of those chemicals can cause real harm.[54]

In particular, the US Academy of Pediatrics warned in a public policy statement that common preservatives and additives which are thought harmless to adults may be harmful to children; it lists a string of substances of concern. 'Children may be particularly susceptible to the effects of these compounds, given that they have higher relative exposures compared with adults (because of greater dietary intake per pound), their metabolic (i.e. detoxification) systems are still developing, and key organ systems are undergoing substantial changes and maturation that are vulnerable to disruptions,' the Academy stated, adding, 'The potential for endocrine system disruption is of great concern, especially in early life, when developmental programming of organ systems is susceptible to permanent and lifelong disruption.'[55]

The sheer number of chemicals in our food is a crude indicator of the scale of the risk – but tells us little about the impact on public health of constant exposure to microdoses of food chemicals, their capacity to accumulate in the human body over time into toxic or cancer-causing doses, or their capacity to mix with one another to form new, potentially poisonous compounds. Even in the best-managed societies – and that means Europe, mainly – scrutiny tends to focus on one chemical at a time, rather than on the complex mixtures and synergies that inevitably occur within an individual's diet and body as a continual process. We may thus be receiving an artificially reassuring picture, compared to the true extent of both public and personal health risks. However, it must also be strongly stated that many chemicals in the food chain deliver great benefits, especially in minimising food losses or preventing food poisoning, and there is therefore a necessary balance to be struck between benefits and risks. Nevertheless, fewer and fewer consumers are reassured and until industry is more transparent about food additives, it is unlikely they will be.[56]

This concern is expressed in a marked trend among middle-income consumers, especially in Europe but also in North America and Oceania and the more affluent cities of Asia, to buy more organic produce and shop at farmers' fresh markets or even buy from organic producers online for home delivery. Global organic food sales reached $243 billion in 2019 and were on track to attain $352 billion by 2025.[57] Although this represents only about 4 per cent of total global food sales ($7.4 billion in 2020), the market for organic food shows around 10 per cent growth a year, compared with overall food demand growth of 3 per cent. The countries with the largest organic food markets were the USA, Germany and France, while highest per capita consumption was in Denmark, Switzerland and Austria. This constitutes a market signal that no self-respecting food retailer can ignore, and some supermarkets now market their own organic home-brand products in response, as well as sourcing new lines from countries still uncontaminated by Western industrial farming methods.

Despite the many justifications for using man-made chemicals in food, the scientific literature suggests that food companies worldwide are increasingly sensitive to consumer concerns about food chemicals – and are engaged in a global hunt for 'natural' substitutes for synthetic chemicals. Whether this will improve overall food safety, or merely transfer the risk to new classes of chemicals, promoted under the 'natural' label, is not yet clear.

The complexity of the modern food chain makes it almost impossible for the average urban consumer to completely avoid pesticides and other chemicals, even if they mainly eat certified-organic fresh produce. Some of these chemicals will sneak into the diet in chocolates, sweets (candy) and ices, soft drinks, tap water, bottled water, fast food, processed food, snack foods and so on, most of which are nearly impossible to obtain in pesticide-free forms. To help consumers make safer choices, the EWG publishes an annual shopping list of the 'cleanest' and 'dirtiest' (most contaminated) fresh fruits and vegetables on the market. Among the clean produce are avocados, sweetcorn, pineapple and onion.[58] Among the 'dirty dozen' are strawberries, spinach, kale, apples and grapes.[59] The EWG says that 90 per cent of the US apple and strawberry samples it tested contained residues of two or more pesticides.

Europe takes a far stricter view of the presence of pesticides and chemicals in food than does the USA and has outlawed some 880 pesticides that did not meet its criteria for causing no harm to humans or the environment,[60] while permitting only 480 for use on farms. Europe also only permits the use of 400 food additives out of the 3000 permitted in the USA. This means that four-fifths of the chemicals permitted in the US food chain are banned from European food.

Unlike many countries, the EC is thorough in regularly testing its food for chemical contamination and monitoring the results of its restrictions. Its 2020 report stated:

Of the 11,679 samples analysed:

- *58% were found to be without quantifiable levels of residues.*
- *40.6% contained one or more pesticide residues in concentrations ... below or equal to the maximum residue levels (MRLs).*
- *1.4% contained residue concentrations exceeding the MRLs.*[61]

The EC member countries surveyed an additional 91,000 food samples for over 800 chemicals and found only 2.7 per cent breached legal limits. The EC concluded chemicals in its food do not, on the strength of these results, present a serious food risk. However, this judgement applies to food alone – not to the combination with other chemicals in the consumer's home, cosmetics, workplace, car etc.

Recognising that different chemicals may cause similar harms to consumers, and that investigating them one-by-one does not give a true picture, the EC has also pioneered cumulative risk assessment, examining the long-term exposure of consumers to combinations of different food chain chemicals. Still in its early stages, its 2020 report found 'consumer risk from dietary cumulative exposure is, with varying degrees of certainty, below the threshold that triggers regulatory action'.[62]

The same, however, cannot be said for other countries. China, for instance, has been in the world headlines for a string of food safety issues,[63] highlighted in the notorious 2008 scandal in which baby formula was adulterated with the plastic melamine, leading to the deaths of six babies and the poisoning of another 54,000. In the ensuing crackdown, two offenders were shot and two more received life sentences.[64] In a food industry as vast as China's, scope for deliberate and accidental chemical contamination of food is large. A study in twenty Chinese cities found 'four fifths (80%) of its respondents were not happy with food safety in China, with food companies (60%) receiving most of the blame'.[65] The government responded to this growing public concern with an extensive overhaul of its laws, tougher penalties, and the restructuring of its controlling authority as the National Medical Products Administration in 2018, responsible for the safety management of food, drugs, cosmetics and medical devices. A rise in prosecutions in recent years also points to a firmer stance by the PRC on food adulteration.

In Asia generally food scandals abound, as regulators scramble to keep up: 'In Asia, not a week goes by without another new food safety scandal making headlines, from contamination and food poisoning, to mislabeling and product recalls.'[66] Here too, consumer pressure for safer, cleaner food, more truthful labelling and fewer chemical ingredients is ramping up.

In the USA, the US Food and Drug Administration made over 1200 food recalls from 2017 to 2020, about half of them for biological contamination and the rest for other reasons, often chemical. Italy is the country with the highest number of recalls each year. However, the problem is a worldwide one, as a glance at the Wikipedia listing for 2011–20 shows.[67]

Owing to the globalisation of food and the relentless quest by giant supermarket chains, fast food chains and food processing corporations for ever-cheaper sources of supply of uniform produce – no matter where it comes from or how it is produced – it is probable that the intake of pesticides and industrial chemicals by the consumer is rising. The dangerous paradox is that consumers want food that is cheaper and more appealing to the eye – and this food involves using greater levels of toxic chemistry to grow, process and keep it attractive and edible for days, weeks and months in the food chain. Furthermore, by the way they present and sell food, supermarkets have conditioned consumers to expect cheaper food – and this expectation entails a heavier chemical load in the diet, with all the health risks that implies.

Nobody is safe from the growing contamination of the food supply. Just because you may live in a well-regulated society does not make you safe from pesticide or additive pollution, either local or imported from the developing world, such is the burgeoning river of today's globally traded food, and such is the universal movement and concentration of pesticides up the biological food chain and in air, water, plants, animals and people. Even people who hunt or grow their own food are not safe from chemical exposure: wildlife and urban soils are both increasingly contaminated.

It is important to remember that the universal penetration of man-made chemicals into the food chain has mainly happened in just the last half century. No previous generations were subjected to such a wholesale chemical exposure – although people would have faced natural toxins from plants, food spoilage or heavy metals in soils. However, the toxic exposure of humans through their food for thousands of generations would generally have been quite small – as the human diet itself was far more diverse than today's: humans presently eat fewer than 1 per cent of all the edible plant species on the Planet as food.[68] Humanity is now moving in totally uncharted waters; we are exposing ourselves to many thousands of novel compounds daily which, in most cases, do not occur in nature and to which our bodies therefore have few defences.

Chemicals in food rarely kill people outright (as in the story at the start of this chapter), but they do set consumers up for an ever-expanding list of food intolerances and degenerative health conditions. Among many consumer groups now attempting to grapple with the expanding octopus of a globally contaminated food supply and its effects on themselves and their families, the Food Intolerance Network (FIN), set up by Sue and Howard Dengate in Australia and New Zealand, provides independent scientific advice to consumers and parents globally about chemicals in food (natural as well as man-made), and the diseases and allergies now linked to them.[69]

As with so many modern families, concern about their diet began with parenthood and aiming for the best for their children's health. Sue, a former teacher, recounts:

After the birth of our daughter, we felt as if we had been launched into a nightmare world of severe sleep disturbance, temper outbursts, arguments, oppositional defiance, asthma and other health problems that lasted for over ten years. I had always assumed that I would know if my child was affected by food additives, because I would see a reaction soon after eating – but that's not how it works.

Food reactions are usually delayed by hours or even days and are almost impossible to identify without an elimination diet. I regard myself as a trained observer but I couldn't work out exactly which foods were affecting my kids. However, we never stopped looking for a solution to our problems and after ten years of searching, a dietitian offered us a three week trial of the Royal Prince Alfred Hospital Diagnostic Elimination Diet which turned out to be the magic answer for us.

The RPAH diet avoids additives and is also low in salicylates, amines and glutamates. These naturally occurring food components can cause the same problems as additives in some people, when found within whole foods such as some fruits, and especially when concentrated by processing – for example in highly flavoured products such as strawberry yoghurt. The diet ensures all potential dietary triggers are removed, one by one, to see the results. Sue continues:

We did the RPAH elimination diet as a family, to support our daughter – but we were surprised to find that we all improved. The challenges afterwards showed that we were all affected by some additives and natural food chemicals although often in very different ways ... Between the four of us we experienced every adverse food reaction listed by our food regulators – from rashes and swelling of the skin, asthma and stuffy or runny nose, to irritable bowel symptoms, colic, bloating, diarrhoea, migraines, headaches, lethargy and irritability – and many more. Although we need to live on a restricted diet while at home in Australia, we have been surprised to find that when we go trekking in Nepal, we can eat traditional food in remote Himalayan villages without suffering from the health, learning or behavioural disorders that the Western diet causes us – although even in Nepal foods are rapidly changing.

Sue's partner Howard Dengate is a food scientist by profession – and well qualified to understand the issues: 'It is the craving of Western supermarkets for "immortal food", combined with a flawed food additive approval process, that has caused many of the huge range of problems reported by our members to the Food Intolerance Network,' he explains. 'Today we receive thousands of emails from people all over the world who have found, like us, that changing their diet has a profound effect and want to thank us for making this information available. Among the many success stories we receive are those from people whose children are now happy, healthy and doing well in their journey through life – and from adults who have battled with chronic conditions such as itchy rashes or painful heart symptoms for years, only to find a change of diet is their answer.'

On their website fedup.com.au, the Dengates explain:

> Our food has changed drastically over the last thirty years, and so have food-related problems. Additives are now used in 'healthy' foods such as bread, butter, yoghurt, juice or muesli bars as well as in junk food. Reactions to food additives are related to dose, so the more you eat, the more likely you are to be affected. A British survey found that most consumers underestimate how many additives they eat: the average consumer eats twenty additives per day and most consumers don't know which foods contain additives.

Besides discussing chemicals and their effects, like EWG and many others, the Food Intolerance Network provides consumers with a shopping guide to minimise their exposure to chemical additives. Consumer groups like these are the bow-wave of a burgeoning international movement by citizens to take back control of their own diets and health from profit-driven industrial food and chemical giants. This demonstrates unmistakably that the educated global consumers of the twenty-first century are joining hands and sharing experiences around the globe, and they are no longer willing to swallow processed foods containing anonymous and potentially harmful substances. Nor are they willing to take on trust the assurances of regulators that food is being properly scrutinised and checked. This is a very important and heartening development, to which we will return in Chapter 9.

However, despite its high profile in the media as a source of concern, food represents only a portion of our total exposure to man-made chemicals and pollutants – and augments the thousands of substances we ingest or absorb from the air we breathe, our drinks or the chemicals that contact our skin. The point is that food toxics *cannot* be regarded in isolation from the combined lifelong assault on our bodies, and the reassurance provided by the testing of foodstuffs for single toxic substances is specious, if not actively misleading.

Scientific studies that quantify the full extent and effects of our exposure to chemicals from all sources are critically lacking, although some early attempts are being made. Consequently, the health and safety of the human food supply remains under a large and growing cloud of uncertainty, risk and concern.

Toxic Tap Water, Dubious Drinks

Drinking water has one of the highest profiles of any chemical contamination issue in the world – and is the focus of major clean-up and prevention efforts by most governments and by international aid agencies.

The World Health Organization lists up to 150 substances that present a hazard to human health if found in drinking water.[70] Its Drinking Water Guidelines encourage governments, national and municipal, to take a risk management approach to providing safer water for consumption. It argues this is easier to adopt locally, than a one-size-fits-all global set of rules.

However, with around 350,000 manufactured chemicals, tens of thousands of adventitious substances, and hundreds of new chemicals being developed and released every year, the enormity of the task of keeping the world's drinking water free of toxic chemistry is ballooning, not receding. Driven by consumer fears, the world bottled water industry has exploded, with consumption rising from 150 to 350 billion litres over the second decade of the twenty-first century (Figure 3.2).[71] However, bottled water may not be quite so clean as people like to imagine. A study by German scientists found no fewer than 24,520 different

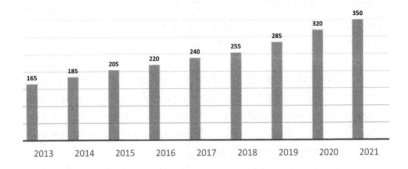

Figure 3.2. Growth in world bottled water sales (in billion dollars) reflects mounting consumer concerns over water safety.

substances in eighteen different brands of bottled water, including hormone disruptors affecting both men and women, plus several new substances previously overlooked by regulators. 'We have shown that antiestrogens and antiandrogens are present in the majority of bottled water products,' they concluded.[72]

A major focus of concern is the substance bisphenol A (or BPA), found in the plastic of many water bottles and babies' bottles. 'BPA is implicated in a variety of health outcomes such as breast and prostate cancer, menstrual irregularities, genital abnormalities in male babies, infertility in men and women, early puberty in girls, and metabolic disorders such as diabetes and obesity. The overall health issues attributed to BPA exposure are complex and controversial,' says one report.[73] BPA is most likely to leach out of plastic bottles made from polycarbonate. These can be identified by checking the Plastic Resin Identification Code (PIDC) on the base of the bottle. Code numbers 1, 3, 4, 6 and 7 are best avoided, as are all plastics with no code number.[74]

The growth in popularity of bottled water reflects the fact that the world is entering a water availability crisis. Two-thirds of the human population already face severe scarcity for at least one month a year.[75] With the population climbing to 10 billion by the 2060s and demand for water rising sharply, both to grow food and to supply cities and mines, the risk of conflict over dwindling resources is also increasing. The volume of fresh water on the Planet is finite, but constantly renewed: in theory, then, there should be no shortages. However, the deadly combination of mismanagement, corruption and pollution of most of the world's accessible surface and groundwater ensures scarcity. By 2040, a US intelligence summary projected, world demand for water will exceed supply, risking instability, increased tensions and state failures.[76]

'The world is facing a water quality challenge due to serious and increasing water pollution, both in developed and developing countries,' says the UN Environment Programme (UNEP). 'This poses a growing risk to public health, food security, biodiversity and other ecosystem services.'[77] Much of the problem of polluted water revolves around 'organic pollution', usually caused by the discharge of untreated domestic or industrial wastewater contaminated with sewage, nutrients, heavy metals and other chemicals. This renders river and lake water both unfit to drink and, often, unable to support inland fisheries.

The challenge of poisonous drinking water is particularly acute in rapidly industrialising countries such as China, India and Brazil. However constant cases of water pollution occur throughout the developing world, affecting up to 2 billion people a year, while, even in developed countries such as the USA and some European states, there are signs of deterioration in local water quality, especially when environmental laws are being relaxed or are flouted by industry.

Half of China's 1.4 billion people cannot access water that is safe to drink or cook with, and the problem is even more acute in rural areas. Water pollution in China was so severe at the outset of the twenty-first century that the World Bank issued a warning that there could be 'catastrophic consequences for future generations'.[78] Chinese media reported in 2016 that 80 per cent of groundwater was undrinkable.[79] At least 70 per cent of lakes and rivers are polluted, and more than half are too polluted for human use. The Yangtze River, China's largest and the world's third-largest river, is inundated with approximately 25 billion tonnes of sewage and industrial waste annually. Algal blooms turn the surface of many Chinese waterways a bright green, but greater problems lurk unseen: groundwater beneath 90 per cent of the PRC's cities is contaminated, mainly with industrial chemicals, forcing them to over-extract from deep aquifers. Rural water pollution has doubled beyond its original government limits mainly due to agricultural, pesticide and fertiliser runoff. Many water sources contain toxic levels of arsenic, fluorine and sulphates which have been linked to China's high rates of liver, stomach and throat cancer. The country's water pollution woes derive mainly from its rapid industrial expansion to meet global demand for cheap goods and the pressure to lift Chinese incomes. Factories, both local and transnational-owned, frequently discharge their wastes into lakes and rivers due to weak environmental rules, lax enforcement and local official corruption. Rural villages located near factories have become known as 'cancer villages' because of their high rates of cancer mortality.[80] Professor Dabo Guan of the University of East Anglia asserted that water pollution is China's greatest environmental issue but feared the public are unaware of its scale. 'People in the cities, they see air pollution every day, so it creates huge pressure from the public. But in the cities, people don't see how bad the water pollution is,' he said.[81]

The situation is no better in another emergent giant, India, where over 70 per cent of surface water is thought to be undrinkable – also due to industrial and sewage discharges from swollen cities, which enter the country's rivers at a rate of 40 million litres a day. Contaminated water is estimated to claim the lives of more than 400,000 Indians a year and cost the economy $7–8 billion.[82]

Even in eco-conscious Europe, the picture is dubious. The European Environment Agency says 45 per cent of the continent's surface waters have yet to achieve 'good chemical status', mainly because of mercury contamination. Nitrates and 160 other pollutants still affect 18 per cent of its groundwater. However, the general trend was one of slowly improving water quality.[83]

In the USA, nearly half of all rivers and streams and over a third of lakes are deemed unfit for swimming, fishing and drinking.[84] American tap water is generally of good quality, with 90 per cent or more passing chemical safety tests – but there is evidence of spreading pollution with toxic substances such as PFAS,[85] arsenic and chromium-6.[86] In 2020 a scientific study reported that 200 million Americans were probably drinking water contaminated by low levels of PFAS.[87]

For years, the 200,000 residents of Flint, Michigan – birthplace of General Motors – were accustomed to chuck their domestic and industrial waste in the local Flint river. This refuse originated from car factories, meatworks, timber yards, paper mills, raw sewage, farm runoff, urban stormwater and leached out of local landfills. Suffering along with America's industrial rustbelt, by the 1980s the population of the city had halved, and so had its revenue, forcing the State Governor to intervene. In 2013, to save money, the decision was taken to switch the city water supply from piped water from Detroit to pumping it out of the local river. Soon after the change, in April 2014, residents began complaining that the water from their taps looked, smelled and tasted foul. Among the contaminants were high levels of lead, which soon led to a doubling, even a tripling of blood lead levels in children. (There is no safe level for lead in drinking water.) This was followed by outbreaks of diseases caused by legionella and *E. coli*. Attempts to cleanse the water with massive doses of chlorine led to a fresh bout of contamination from chlorine byproducts. Efforts by frantic citizens to have an emergency declared were rebuffed by federal officials, leading them to launch a series of legal actions. These eventually led to court orders to the City to deliver bottled water to affected residents and replace thousands of unsafe old lead water pipes, but five years later the situation still dragged on.[88] Like Minamata in Japan, Bhopal in India and Tianjin in China, the consequences of Flint's toxic water will echo for years, and in some cases for lifetimes.

In microcosm, the saga of Flint epitomises the issues now facing residents of most of the world's megacities as yesterday's chemical legacy returns to haunt their health and their future, and as fewer and fewer water sources are found to be free of humanity's Planet-wide pollution. This raises a secondary problem for water quality: the increasing use of chemical treatment to cleanse polluted water is itself delivering new forms of chemical pollution: even 'clean' tap water carries health risks. When chlorine and bromine are used to disinfect domestic water of microorganisms, they react with natural substances present in the water from the decay of organic matter, such as leaf mould. This chemical reaction produces a family of toxic byproducts known as trihalomethanes (THMs).[89] These have been linked in medical research with liver and kidney problems, bladder cancer, retarded foetal growth and spontaneous abortions.[90] It is another case of the two-edged chemical sword.

Around twenty-five countries worldwide add fluoride to their drinking water to prevent tooth decay in the 500 million people who drink it. Many countries also drink groundwater that is naturally rich in fluorine. Since it was first introduced in the 1940s, artificial fluoridation has been the subject of public controversy, both over its health impacts and its dental benefits. Despite numerous studies and inquiries, the situation for both remains unclear. It is just one more substance added to the compound chemicalisation of the human environment.

Avoiding tap water in favour of manufactured drinks does not avoid the problem of contamination, as these drinks have many chemicals added to those in the water itself. 'Soft' drinks are in fact mostly a chemical cocktail consisting of water (90–99%), sweeteners (sugar or various chemical substitutes), carbon dioxide, food acid, flavourings, dyes, chemical preservatives, antioxidants and foaming agents. In some cases these ingredients may react together to produce highly toxic substances, such as cancer-causing benzene. Converted to disease potential, soft drink ingredients have been separately and collectively linked to increased risk of tooth decay, obesity, neurotoxicity, hyperactivity, kidney, liver and gut problems, skin and eye diseases, heart attacks, stroke, cancers and gene damage. A European study of nearly half a million people linked soft drink consumption to earlier death from all causes. Consumption of artificially sweetened soft drinks was linked with deaths from circulatory diseases, and sugar-sweetened soft drinks were associated with deaths from digestive diseases.[91] A typical can of soft drink contains seven to ten teaspoonfuls of sugar, according to Harvard School of Public Health. Drinking just two cans a day increases your risk of early death by heart attack by 31 per cent – and every additional can by 10 per cent more.[92]

It is the cumulative exposure to all the various chemicals in soft drinks and the human diet that is the real concern. 'Soft drinks consumption is still a controversial issue for public health and public policy ... there is a strong body of evidence to support the existence of health risks associated, especially, with overconsumption and with certain artificial colorings and preservatives,' one study concluded, warning that consumers are still blithely unaware of most of the risks.[93]

Even those beloved beverages, coffee and tea, are not entirely harmless, new science reveals. Both may include many chemicals, including pesticides (some of them banned) used on the developing country farms that grew them, heavy metals, acrylamide, fungal poisons and, of course, their main natural ingredient, caffeine. The science generally indicates that neither drink is dangerous if consumed in moderation and if made without excessive boiling of the beans/leaves in water. Indeed, they probably have health benefits and may help reduce the risk of some diseases. However, as with many processed foods, more chemicals are now entering the natural product during growing, processing and packaging. Caffeine itself is a poison in large doses, and overconsumption should be avoided.[94]

Ageing Accumulators

The older you get, the more polluted you are. The rate of chemical uptake by the human body may increase with age, recent studies suggest. This is because older skin is thinner and more permeable to toxins, while the kidneys, liver and blood supply are less efficient at breaking them down and removing them from the body.[95]

Nowadays, too, there is a trend towards obesity in older people, and fat is where many persistent substances (such as POPs) accumulate. The longer these remain in the body and the more they concentrate, the greater the likelihood of damage to your health. This suggests that the incidence of chemical poisoning and related diseases is likely to rise as an unantici-pated burden of the ageing, increasingly overweight population. Accumulated toxins in older people may also interact with other diseases or effects of ageing, leading to a weakened immune system, which in turn may increase the risk of cancer or infec-tion, for example.

Postmortem Pollution

Death itself marks no boundary to the processes by which we humans release and recirculate toxic substances. 'Almost all cemeteries have some potential for pollution,' comments a world-first study of the emissions from burial grounds, carried out in Australia.[96] This found that most of the long-lived chem-ical substances contained within a corpse at the time of death are re-released into groundwater within ten years of its burial, putting them 'back into circulation' and so exposing future generations to historical toxins. While the most obviously risky discharges from burials are bacteria and viruses, there are reports of more subtle and long-lasting releases of heavy metals and persistent organic pollutants being found in groundwater downstream of burial grounds – groundwater that, not infre-quently, ends up as urban drinking water. Humans, as long-lived, top bioaccumulators, store more of these substances in their bones and fat than do other animals – and give them back to future generations in a process that has all the hallmarks of a Pharaonic curse.

While cemeteries are not to be compared with, say, landfills or contaminated industrial sites as a concentrated source of future health risks, these findings still demonstrate the long life of some forms of toxic pollution, their capacity to recycle within the Earth system and future human population via water supplies and food chains. This raises the disturbing reflection that chemical pollution today, accumulating generation upon generation, is a morbid heirloom. Its toxic consequences reach into the distant future and to our descendants yet unborn.

The Toxic Toll

Pollution is the largest environmental cause of disease and premature death in the world today. Diseases caused by pollution were responsible for an estimated 9 million premature deaths in 2015 — 16% of all deaths worldwide — three times more deaths than from AIDS, tuberculosis, and malaria combined and 15 times more than from all wars and other forms of violence. In the most severely affected countries, pollution-related disease is responsible for more than one death in four.

So declared *The Lancet* Commission on Pollution and Health in 2018, adding that pollution also kills the poor, the sick, the urban and the vulnerable in disproportion.

Chemical pollution is a great and growing global problem. The effects of chemical pollution on human health are poorly defined and its contribution to the global burden of disease is almost certainly underestimated. More than 140,000 new chemicals and pesticides have been synthesised since 1950. Of these materials, the 5000 that are produced in greatest volume have become widely dispersed in the environment and are responsible for nearly universal human exposure. Fewer than half of these high-production volume chemicals have undergone any testing for safety or toxicity, and rigorous pre-market evaluation of new chemicals has become mandatory in only the past decade and in only a few high-income countries. The result is that chemicals and pesticides whose effects on human health and the environment were never examined have repeatedly been responsible for episodes of disease, death, and environmental degradation.[97]

The Lancet freely acknowledged its number of 9 million deaths due to chemical poisoning is an underestimate — and probably a

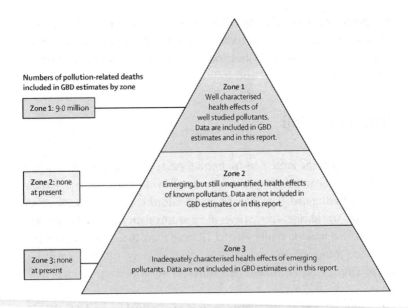

Figure 3.3. The 'Pollutome' or combined health impact of our chemical emissions on humanity (GBD: global burden of disease). Source: *The Lancet*, 2018.

very large underestimate – as the diagram in Figure 3.3 makes clear.

For example, the transformation of healthy human cells into cancer cells is triggered by exposure to certain chemicals in 30 to 50 per cent of cases, according to various estimates, though the disease often takes years to appear. With 10 million cancer deaths worldwide, this means there may be 3–5 million *additional* fatalities a year, attributable to chemical poisoning. The same applies to deaths from chemically related obesity, heart disease, depression (suicide) and other conditions. Between them, these could double the 9 million fatalities estimated in Zone 1. Quantifying the full impact of anthropogenic chemicals on humanity is a task that bears comparison with the Apollo moonshot. Nonetheless, some are attempting it, one of them being Professor Howard Hu of Washington University, who aims to fill in the blanks in Zones 2 and 3 in *The Lancet* pyramid by including pollution in the ongoing *Global Burden of Disease* (GBD) study.[98]

'The good news is that, despite the great magnitude of pollution and current gaps in knowledge about its effects on human health and the environment, pollution can be prevented. Pollution is not the inevitable consequence of economic development,' *The Lancet* authors stated, adding, 'Contrary to the oft-repeated claim that pollution control stifles economic growth, pollution prevention has, in fact, been shown repeatedly to be highly cost-effective.'

The bottom line here is that millions of early deaths are, potentially, avoidable by eliminating the fatal sources of the pollution.[99]

The real question is: do we wish to prevent them?

4 DIABOLIC COCKTAIL

'All things are poisons, for there is nothing without poisonous qualities. It is only the dose which makes a thing poison.'

Paracelsus

For eight days, five San Francisco families were tested by researchers for the presence of common hormone-disrupting chemicals used in food packaging. The families ate their normal diet for five days, and a healthy fresh food diet for the remaining three, then had their urine tested. 'When they compared urine samples before and after the diet, the scientists were stunned to see what a difference a few days could make,' *The Washington Post* reported.[1] The scientists found that substances such as BPA and phthalates were reduced by half to two-thirds and, in some cases, over 90 per cent, just by switching to fresh food.[2]

The study was revealing in three ways. Firstly, it showed just how stuff that can mess with your body and ability to have children is getting into a normal diet – and there are now many reports that show this. Secondly, it confirmed that we are dealing with not just single chemicals, but pernicious mixtures. And thirdly, it proved beyond doubt the power of the consumer to clean up their own food supply and improve their health outlook by voting with their dollar and avoiding 'industrial food'.

There could be anything up to 16,500 different chemicals in the modern food chain today that simply weren't a part of your grandparents' diet. It isn't just a case of a few nasty ones, but of thousands upon thousands of substances, their breakdown products and their recombinations which now menace you, your life and your loved ones.

Figure 4.1. Portrait of Paracelsus by Augustin Hirschvogel, 1538. Source: Wellcome Library, London.

For years, the chemical industry has employed a mediaeval alchemist to defend itself against its critics, and to justify its continued chemicalisation of our food, our air, our water, our bodies and our lives. His name was Philippus Aureolus Theophrastus Bombastus von Hohenheim and he lived in Switzerland from 1493 to 1541 and was known, for short, as Paracelsus (Figure 4.1). He is most famous for his dictum '*sola dosis facit venenum*' – the dose alone makes the poison – by which he meant that all substances, even water, can be deadly if you take in too much of them. It's a question of quantity. Nobody really argues with that. Five hundred years ago, it was pretty good 'science'.

However, Paracelsus' aphorism has since been hijacked by the modern chemical industry to signify that, so long as it is below the accepted residue limit in your food or drink, a single pesticide or other substance is therefore harmless – and you ought not to worry about it. This claim is wrong on so many grounds it is hard to know where to begin – but here are a few:

- For many, if not most chemicals, the 'harms' are largely unknown.

- No item of food or drink has only one man-made chemical or pesticide in it.
- Chemical levels that have tested safe in adults may be unsafe in children and infants.
- Some chemicals may affect women more than men, since maximum dose rates have been calculated based on average male body size.
- The 'untested chemical assumption'[3] means that chemicals which have never been scientifically tested and reported on are presumed safe for regulatory purposes.
- Some hormone disruptors appear to cause a larger effect with a lower dose.
- It does not cover the new chemicals that form from the breakdown of those in the diet.
- It takes no account of constant, frequent exposure to the same substance over time, in varying doses.
- It takes no account of the tendency of some substances to accumulate in the human body, reaching potentially toxic levels over time.
- It takes no account of the impact of consuming several completely different chemicals which all lead to the same harm or which damage the same cell types.
- It takes no account of the toxic synergies of thousands of substances, both natural and synthetic, blending together in the diet, short term or over a lifetime, the byproducts they may form, or the cascades of health-harm they may trigger.
- It takes no account of the phenomenal increase in the use of chemicals in all aspects of modern daily life.

Typical of the industry's excuses are statements such as 'Just because a chemical is present does not mean it is harmful in the amount present,' uttered by a group calling itself *Sense About Science*, which then went on to claim that apples, pears, potatoes and zucchini all contain natural toxins in small amounts, yet are safe to eat.[4] Superficially, the statement is accurate – but it is deliberately disingenuous; its purpose is to reassure the public that there is 'nothing to see here', in order to protect industry profits. This flies in the face of tens of thousands of scientific papers raising concerns about substances in the modern food supply, water and home – and linking them with a host of modern diseases. And it completely ignores the emerging science of mixtures.

Just why the modern chemical industry needs to hide behind mediaeval alchemy – the study of materials from a time before humans even knew that molecules existed – is an interesting question for historians and philosophers of ethics. The answer undoubtedly is to be found in economics: it's all about money. For humanity at large, however, it is a life and death matter, as *The Lancet* Commission has made clear.

The presence of thousands of chemicals in food packaging and containers as well as in the food itself illustrates in part how these man-made substances now surround us, in forms and in combinations that are often hard if not impossible for most people to avoid without radical change to their lives and consumer habits. You may choose fresh, organic produce in your local supermarket – but did it come in a plastic wrapper? You may feed your baby perfectly clean and harmless milk formula – but was it from a plastic feeding bottle or cup? Packaging is important for food preservation and many problems such as food spoilage, poisoning and even starvation which would occur without it, so it isn't a matter of getting rid of the packaging – but rather, of finding safer alternative materials.

All this highlights a sleeping giant of an issue – that all humans are now affected, lifelong, not just by individual chemicals of concern, but by an enormous cocktail of substances emanating from numerous apparently innocuous sources in our daily lives. These act in concert with one another and affect our health in ways that are, as yet, poorly understood – but which are starting to become clearer.

The sheer magnitude of the problem puts it alongside climate change in terms of its impact on the human future – and ahead of climate change in terms of its impact on our immediate personal wellbeing and survival. The full scale of this impact is not yet acknowledged by science, nor by governments, nor by industry, nor by the community. While consumers are smart enough to be concerned, they have – as yet – little appreciation of the sheer size of the beast.

Whether it is a blend of toxins and heavy metals leaching from a former industrial site and getting into our air, water or food, hydrocarbons entering groundwater from a former petrol station or fuel dump, a witches' brew of poisonous substances from an old urban landfill dissolved in our drinking water, a cocktail of pesticides, preservatives, chemical dyes and additives in the daily diet or a whiff of volatiles from cosmetics, furnishings, clothing and vehicles, modern humans are constantly assailed by thousands of different man-made or -emitted chemicals every day – indeed almost every moment – of their lives, in ways that previous generations simply were not.

Despite the magnitude of this assault, the health effects of this battery of substances, in both the short and long term, remain largely uncertain. This is despite the fact that chemical mixtures have been a focus of public health concern for more than three decades – the USA, for instance, first began to develop guidelines for assessing mixtures in 1984.[5]

Over a third of a century later, Elina Drakvik and colleagues of Sweden's Karolinska Institute lamented the glacially slow rate of progress in dealing with the mixture threat: 'The number of anthropogenic chemicals, manufactured, byproducts, metabolites and abiotically formed transformation products, counts to hundreds of thousands, at present. Thus, humans and wildlife are exposed to complex mixtures, never one chemical at a time and rarely with only one dominating effect. Hence there is an urgent need to develop strategies on how exposure to multiple hazardous chemicals and the combination of their effects can be assessed.'[6]

A similar perspective was advanced by Harvard School of Public Health: 'There is a critical need to understand the health effects of chemical mixtures, to protect the public's health. It is well accepted that humans are routinely exposed to multiple chemicals simultaneously or sequentially, and there is evidence that the toxicity of individual chemicals depends on their interactions with other chemicals.'[7] The review concluded that the health impacts of mixtures were often different from those of chemicals ingested singly.

Put simply, you can't just take a mixture of noxious chemicals and put it into a person to see what happens: the experiment would be barred on ethical grounds. Unless it happens to be the whole human species, in which case it is apparently OK. The real difficulty, as with tobacco, is connecting a specific disease with a multitude of potential chemical triggers, known and unknown. It is also defining which mixture – air pollution, diet, water, domestic or workplace – is most responsible for a particular cancer or other health outcome.

Meanwhile, studies of the adverse consequences of particular chemical mixtures continue to pile up in the scientific literature. Here are just a few examples:

1. Danish scientists found that lab rats suffered deformed penises and other gender problems when exposed to four of the common hormone disruptors found in human food and packaging.[8]

2. Chemicals not known to produce cancer, acting cumulatively in low-dose mixtures on different human tissues, appear to cause cancers.[9]

3. A study found links between mixed endocrine disruptors and autistic behaviour in children.[10]

4. Women exposed to larger mixtures of PFOA, PFHpS and PFAS had a greater risk of miscarriage.[11]

5. Exposure to mixed endocrine-disrupting chemicals in the European population were causing disease and dysfunction across people's lives at a probable cost of hundreds of billions of euros per year to society.[12]

6. Mixtures including air pollution, some metals, several pesticides, some VOCs and phthalates 'showed associations with autism'.[13]

7. The toxic effects of lead, a well-known nerve poison, increase in children exposed to several metals in mixtures.[14]

8. Babies exposed to arsenic, manganese and lead while in the womb showed signs of complex nerve poisoning affecting their development in early life.[15]

9. Low-dose chemical mixtures may help to promote tumour angiogenesis in cancers.[16]

10. Higher levels of metals in air pollution, specifically mercury, cadmium and lead, are associated with a higher risk of breast cancer in older women.[17]

11. Industrial workers in factories may be routinely exposed to complex mixtures of carcinogens, depending on the nature of their work.[18]

A particularly significant piece of research, carried out by a team of European scientists, measured the exposure levels of forty-one chemicals – including BPA, phthalates and pesticides – in the blood and urine of more than 2000 pregnant women in Sweden. When the chemicals were tested one by one in human cells and in lab animals, there seemed little to be concerned about. However, when mixtures were tested on animals at the levels found in the women, sexual development and metabolism were both affected. 'We saw effects in many of our (animal) models at levels comparable to what we could measure in the pregnant women,' said Dr Joelle Ruegg, a molecular toxicologist at the Karolinska Institute in Stockholm. Based on levels in the pregnant women, 11 per cent of the unborn children were at risk of problems with their sexual development due to exposure to common chemical mixtures. Had the chemicals been examined one at a time, this figure would only have been 1 per cent. The mixture had resulted in more than a tenfold increase in risk.[19]

We are, every one of us, the 'laboratory rats' in a vast worldwide chemistry experiment involving an immense cocktail of substances, over which we have neither say nor freedom of action.

It is an uncontrolled global experiment that defies the very ethics which outlaw the scientific testing of mixture toxicity in humans.

It is an experiment whose parameters have been defined for us by 200 giant fossil fuel, petrochemical and pharmaceutical corporations operating globally, with the tacit approval of governments, paying lip-service to inadequate laws in some countries and ignoring them in others.

It is an experiment now known to bring death to millions.

Mixture Troubles

'Humans and all other organisms are typically exposed to multi-component chemical mixtures, present in the surrounding environmental media (water, air, soil), in food or in consumer products,' commented Professor Andreas Kortenkamp and co-authors in the EU's pioneering State of the Art Report on Mixture Toxicity. 'However, with a few exceptions, chemical risk assessment considers the effects of single substances in isolation, an approach that is only justified if the exposure to mixtures does not bear the risk of an increased toxicity.' The report points out chemical mixtures can arise in several ways:

- As a result of substances in which several chemicals are already mixed.
- In manufactured products combining several different chemicals.
- Through interaction during production, processing, transport, consumption or recycling.
- By chemicals becoming mingled in air, water, soil, plants and animals, food and in human tissues, via many different pathways.[20]

To put it plainly, modern chemical risk assessment, however well intentioned and thorough it may at times be, tends to gloss over the big picture, especially at global scale. As David Carpenter and colleagues from New York State University originally explained: 'In reality, most persons are exposed to many chemicals, not just one or two, and therefore the effects of a chemical mixture are extremely complex and may differ for each mixture depending on the chemical composition. This complexity is a major reason why mixtures have not been well studied.'[21]

Add to this the fact that many individual chemical components themselves have been poorly studied: 'Of the tens of thousands of chemicals on the market, only a fraction has been thoroughly evaluated to determine their effects on human health and the environment,' cautioned UNEP.[22] Even in the best-regulated societies, chemical toxicity is habitually investigated one chemical at a time; no assessment of the nature or scale of the risk we face from constant, perpetual exposure to chemical mixtures from all sources is available.

Although many mixture-toxicity experiments have been carried out in the laboratory – chiefly by exposing simple organisms such as worms, plant seeds and microbes to carefully chosen mixtures – most of these have so far assumed that the increase in toxic risk is arithmetic, adding the risk factor for one chemical to that of another to obtain a total estimate for the mixture as a whole. However, recent research suggests the risks when two or more chemicals are mixed are greater; toxicity may instead be non-linear, meaning that different chemicals not only add toxicity on their own behalf but may also combine with others to produce a much larger overall toxic dose or form new toxic compounds. Underlining this, the authors of the EU mixture-toxicity report state:

> There is strong evidence that chemicals with common specific modes of action work together to produce combination effects that are larger than the effects of each mixture component applied singly. Fewer studies have been conducted with mixtures composed of chemicals with diverse modes of action, with results clearly pointing in the same direction: the effects of such mixtures are also higher than those of the individual components.
>
> There is a consensus in the field of mixture toxicology that the customary chemical-by-chemical approach to risk assessment might be too simplistic. It is in danger of underestimating the risk of chemicals to human health and to the environment.[23]

Until recently, too, there was an assumption that if every chemical of concern in a mixture were to be kept below its particular threshold of known harm, then the overall mixture would be safe. Tests in microbes, water fleas, fish and human cells have debunked this idea, the EU mixture report states. 'Hence, any concentration of any compound needs to be considered because it adds to the mixture concentration.'

'There is decisive evidence that mixtures composed of chemicals with diverse modes of action also exhibit mixture effects when each component is present at doses equal to, or below points of departure (harm thresholds),' the European scientists affirmed.

Subsequently, the EU Directorate-General for Health and Consumers has said:

> *Since humans and their environments are exposed to a wide variety of substances, there is increasing concern in the general public about the potential adverse effects of the interactions between those substances when present simultaneously in a mixture. In view of the almost infinite number of possible combinations of chemicals to which humans and environmental species are exposed, some form of initial filter to allow a focus on mixtures of potential concern is necessary ... A major knowledge gap at the present time is the lack of exposure information and the rather limited number of chemicals for which there is sufficient information on their mode of action.*

In short, as the European health authorities caution, we simply do not comprehend the danger we may be in, or the risks we run as members of a complex industrial consumer society. The information we need to help us to make safe and responsible choices is currently, at best, fragmentary – and it tends to suggest that by avoiding particular products, or by changing our behaviour in certain ways, we can eliminate or mitigate the risk to ourselves. However, the emerging evidence indicates this may be delusory.

Only by reducing the total number and overall toxicity of *all* the man-made chemicals present in our world and daily lives can we limit or reduce our risk.

Billions of Mixtures

Given the many billions of possible mixtures that the 350,000 man-made chemicals can form – not to mention the countless other natural or adventitious substances created or found in society's waste streams, eroded soil, mineral discharges and fossil fuel emissions – the task of assessing every possible mixture for its effect on human health, chemical by chemical, might well appear insuperable. This is certainly an excuse oft trotted out to explain why governments or industry are doing so little, on any meaningful scale, about it.[24]

However, the authors of the EU report consider the task is not impossible: 'There is strong evidence that it is possible to predict the toxicity of chemical mixtures with reasonable accuracy and precision. There is no need for the experimental testing of each and every conceivable mixture, which would indeed make risk assessment unmanageable,' they say. This applies especially to mixtures of known chemical composition (for example, most mixtures found in our food) where both dose rates and the independent action of substances in the mix can be considered together. Where ingredients are incompletely known (for example, as in the case of an old industrial sludge underlying city housing), the problem remains challenging.[25]

With the health and lives of tens of millions of people at stake, there is no ethical justification for ignoring this challenge on the mere grounds that it is difficult or expensive. The advent of sophisticated chemical and genetic modelling and supercomputers which allow researchers to quickly perform vast calculations enable society to make a reasonably educated prediction about the mixtures that pose the greatest potential hazard, and the groups of people who are most at risk.[26]

Yet such scientific approaches are not being adopted globally in a concerted manner, or with any sense of urgency. This seems strange, given that the risk is universal: no person, nation or community is exempt from it. For some reason, society at large continues to be deceived about it.

Over a lifetime, each of us is exposed to highly complex mixtures of many thousands of chemicals, generally in small doses, sometimes for short periods, sometimes constantly over many years, and sometimes in large pulses. Some of these substances we flush from our bodies readily, from some we sustain transient damage, and some (such as certain heavy metals and persistent chemicals) we bioaccumulate over many years, building up our own potentially toxic dose over repeated exposures to many non-toxic doses. However, the extent to which chemical mixtures attack or undermine our health, and that of the environment, over time remain scientific unknowns.

America's CDC has a Chemical Mixtures Program that is 'mandated to determine the health impact of exposure to combinations of chemicals', but which immediately narrows its main focus to mixtures most likely to be found at hazardous waste sites.[27] These, as we have seen, are only a fraction of the possible sources of chemical mixtures in the average person's daily life. Furthermore, such an approach conforms to the outdated notion that contamination by mixtures is a local, rather than a universal, problem.

The complexity of this challenge led organisations such as Australia's Cooperative Research Centre for Contamination Assessment and Remediation of the Environment (CRC CARE) and others to investigate ways to reveal the overall state of toxification of an individual from all sources, rather than attempting to separate and attribute their ailments to particular substances. This approach uses a range of indicators, including levels of certain proteins in the blood, immune system status, measures of genetic damage and so on. Although it does not help track down the specific poisons at fault, this approach – if successful – at least has the virtue of disclosing how poisoned an individual may be as a result of their total chemical exposure over time. If such tests were used for general screening and combined with epidemiological studies, it may then be possible to narrow down the main sources of toxicity and identify which are doing the most harm. Furthermore, the study of many different individuals may enable a society-wide estimate which may reveal the broad patterns of toxification in various groups of exposed people and, hence, the likely main sources of their exposure.

Mixtures and Health

The effect of chemical mixtures on human health is, as might be expected, as complex and murky as the mixtures themselves.

Many chemicals appear to act on most of the cell types in our body. They may invade the cell and scramble its functions, they may reprogram what its genes do, causing it to affect other cells, or they may kill it. They may simply block the cell's ability to receive the normal 'healthy' chemicals we rely on for life, such as calcium or hormonal signals.

Where there are several mixed chemicals at work, they can have a wide range of different and interacting effects, making accurate diagnosis especially problematic for doctors grappling with a range of symptoms. For example, each chemical may have quite different actions on the kidney, the liver and the brain, each with a different disease effect – sometimes even a benign one – or else there may be a puzzling combination of benign and malignant effects. Such experiences are well known with medical drugs and their side-effects, and undoubtedly apply to chemicals in general.

Nevertheless, a growing number of studies have revealed that a wide range of diseases not previously thought to be caused by 'environmental factors' – such as heart disease, diabetes, bone and joint disorders – are in fact linked to exposure to man-made contaminant mixtures. Precisely what the link is, or how much it is influenced by mixtures of chemicals as opposed to single substances, remains hard to pinpoint.

Professor David O. Carpenter and colleagues have identified the following broad categories of human disease as being partly due to the effect of chemical mixtures:

- Developmental disorders: disorders that occur when exposure to chemical mixtures in pregnancy, infancy or childhood leads to lifelong mental, physical and reproductive disorders.
- Neurobehavioural abnormalities: including a lifelong deficit in intelligence caused by exposure to toxins such as lead and PCBs in the early years; nerve disorders brought on by mercury exposure; intelligence and co-ordination problems due to pesticide exposure; reduced learning capacity caused by mixtures containing PCBs, dioxins, furans, lead and mercury.

- Sexual disruption: caused by chemicals which mimic or alter the natural functioning of sex hormones in the very young. These result in feminisation of males; masculinisation of females; reproductive deformities; diseases of the womb and prostate; infertility in both males and females; reproductive cancers; and possibly, changes in sexual preference.
- Neurodegeneration: includes a range of diseases that result in the death of nerve and brain cells, such as Parkinson's, Alzheimer's and amyotrophic lateral sclerosis (ALS). While the links are still largely speculative owing to the lapse of time, more scientists are now attributing these mature-onset diseases to the early exposure to toxic chemical mixtures.
- Cancers: genetic factors alone are thought by scientists to be responsible for no more than 5 per cent of cancers. The rest have an external trigger, although there may be a genetic predisposition that makes the individual more susceptible to the particular cancer. The biggest 'killer mixture' of chemicals is delivered in cigarette smoke (accounting for 30 per cent of all cancer deaths), but there are plenty of other suspect chemicals, including oestrogens, organochlorine pesticides, PCBs and polyaromatic hydrocarbons, which have all, for example, been linked with breast cancer.
- Heart disease: while heart disease and stroke are most commonly linked to fat and diet, they are also quite strongly connected in the scientific literature to exposure to chemical mixtures, containing for example heavy metals, certain pesticides and cigarette smoke. The theory is that these chemicals not only damage cell function, but also cause direct cell damage.[28]

How chemical mixtures actually cause these diseases is a matter of extraordinary biochemical complexity, which is still in the process of being unravelled by science. Because progress has been slow in disentangling the many and varied effects of chemical mixtures on human health, progress in regulating or controlling it has been equally slow.

It is vital to understand that this does *not* mean that the risks are small – only that they are extremely difficult to trace and quantify. In other words, we are almost certainly underestimating the dangers, probably by a large margin.

Anyone living in today's world is subjected to an unprecedented bath of man-made or -emitted chemical mixtures, head-to-toe, every day of our lives. While some scientists argue that the fact that more people are alive and living longer bespeaks our ability to adapt to the toxic flood, others are not so sure, fearing that worse may be in store for the health of both individuals and society as a whole in coming years, unless the weight of our overall chemical burden is somehow reduced. The 'longer lives' argument is, in any case, a distraction: it only reflects causes of mortality for people in the past – not new and emerging lethal impacts on people now living. Those have yet to come.

At present there is no practical way to contain the impact of chemical mixtures – either on humans or on the natural environment. Since these substances are emitted from multiple sources – not only the chemical industry itself but also mining, power generation, manufacturing, transport, construction, pharmaceuticals, waste disposal, food processing, cosmetics and so on – and reach people by multiple pathways, it has so far proved impossible to regulate or control their production, dispersal and recombination in mixtures. If the pattern of behaviour established by big tobacco companies is pursued by fossil fuel and petrochemical companies – as is indeed the case[29] – an industry that deliberately or unintentionally produces toxic substances will as a general rule begin by denying its contribution, then hire its own scientists in a bid to disprove it, then blame every other industry for the impact of mixtures before at some point reluctantly conceding its own contribution, and agreeing to limit it. This process typically spans decades, stalling progress and resulting in countless unnecessary deaths.

The task for government in pursuing industries whose products may in some cases only be harmful when interacting with other chemicals in mixtures is even harder, as the trail of responsibility is more difficult to trace and assign. Yet tens of millions of people are suffering and dying – and hundreds of millions more will die in coming decades – as a consequence of

the general release of these substances, their combined interaction and our inability to assess or prevent the damage they
are causing.

For tens of millions of people to die needlessly for any preventable reason is an affront to civilisation. It is a basic moral
principle that such preventable deaths should not be tolerated
by society any more than other forms of manslaughter. Yet, in
the case of chemicals, we have inexplicably forsaken our duty of
care – or chosen to overlook it. The millions of infants and
children who will have their young lives affected by exposure
to these noxious mixtures before even attaining an age where
they can refuse, reject or seek to avoid them represent a grievous
delinquency on the part of society. We lust for the goods and
benefits of modern consumerism. Yet we shun the moral responsibility for the harm it inflicts.

'Environmental' Chemicals

One of the sneaky tricks employed to keep the public in the dark
about the risks of chemicals and their mixtures is use of the
term 'environmental chemical'.[30]

Anyone who has read any of the scientific literature will be
aware this phrase is used, almost universally, to describe chemicals *in the human living environment*, whether natural or man-
made. The public, unfamiliar with the arcane practice of science
hijacking familiar terms and applying their meaning to something very specific in a particular field, will generally interpret it
to mean a chemical that occurs in nature (which is 'the environment', to most people). And, to the public, natural = good, safe,
clean, wholesome – as cynical advertising agencies have
long known.

By calling something man-made an 'environmental chemical',
the experts who use this term are inviting the public to miscon-
strue the toxic risk they face. In this case, science is behaving as
unscrupulously as advertising.

There can be little doubt that the term has come into common
use in science through the efforts of the chemical industry

which, like big tobacco, would prefer everyone to believe its products harmless. By calling pesticides 'environmental chemicals' they invite the inference that they are safe and wholesome, instead of the opposite. Or that furniture flame retardants are 'environmental' even when found in deep-sea squid.

The point is accentuated by the discipline of *environmental chemistry*, which is 'the scientific study of the chemical and biochemical phenomena that occur in natural places' and hence a correct use of the term.[31] Indeed, one of the very goals of this profession is to separate out the effects of natural and man-made chemistry in the natural world.

To call air pollution, BPA, dioxins, benzene, PFOS, food dyes, 'gender benders' and other toxicological assaults on human health 'environmental chemicals' is not only disingenuous, it is dangerous – because many people will assume such things to be safe purely on the reassurance of the word 'environmental'.

The dictionary defines 'environment' in two main senses: (1) The living environment that surrounds you; and (2) The natural world. The use of the term 'environmental chemical' by science furthers public confusion over the risks they face by blurring the second meaning with the first and forgetting to mention that its use of 'environment' also includes the artificial environment made by humans.

The purpose of science is to clarify, not confuse, the world we live in. Good science is careful to avoid ambiguity or scope for misinterpretation. However, this is not the case with global poisoning. Here, it appears, even the scientists who are working to clarify, understand and prevent it have been roped into the chemical industry's clever game of branding its outputs in a way the public assumes means 'natural', ergo 'safe'.

Many jurisdictions, such as the European Community, have strict laws around the branding of food products as 'natural' because of the scope for deception by unscrupulous companies and for misunderstanding by consumers. It is time for science to apply equally firm rules to chemistry.

From here on, those who use the term 'environmental chemical' must confine it to meaning a substance that occurs in the natural world, one that is not made or generated by humans. Or else risk being a party to one of the most deadly coverups in human history.

5 UNSEEN RISKS

'I wasted time, and now doth time waste me.'
William Shakespeare, *Richard II*

Amid the acrid reek and choking smog from open fires and iron cauldrons, in a place where it seemed half the world's computer scrap was being rendered down just to recover a few miserable metal traces, a young Chinese scientist was studiously gleaning information to help her team understand the health problems now evident in the workers and children enslaved in this New Age Inferno. As she spoke with some of them, a group of rough-looking men approached, demanding to know what she was doing. When she tried to explain her intention of gathering information for her team's health research, the knives came out. One of the men slashed at her, opening an ugly wound on her hand with his knife, as she fought to deflect the blow. Bleeding and terrified, she and her colleague fled headlong down grim shanty streets, past hovels and yards piled high with high-tech detritus, pursued by their angry assailants. It was just another grimy, vicious day in Guiyu, south China, a place that had come to symbolise the ugly, toxic face of the IT revolution.

In 2005, the *South China Morning Post* wrote of Guiyu: '(The) landscape ... varies from filthy to apocalyptic. In small work-shops and yards and in the open countryside workers dismember the detritus of modernisation. Armed mostly with small hand tools they take apart old computers, monitors, printers, video and DVD players, photocopying machines, telephones and phone chargers, music speakers, car batteries and micro-wave ovens.'[1]

Fifteen years on there were signs of healing. Thousands of polluting streetside workshops had shut down and others had been herded into the Guiyu Recycling Economy Industrial Park in a bid by Chinese authorities to clean up the hellscape and crack down on pollution. The air was cleaner, workers safer – but also poorer.[2]

In 2017–18, China, which had been handling up to 70 per cent of the world's e-trash, announced it would ban the importation of thirty-two categories of solid waste, including electronic goods and plastics, for reprocessing. In 2020 the PRC affirmed its goal was 'gradually realizing zero import of solid waste'.[3] The bans caused a major shock, throwing the problem back into the laps of the main offenders – Europe and North America: the USA alone had been shipping 4000 containers of waste *a day* into the PRC. A desperate attempt to offload their trash on Thailand and Vietnam failed when both countries imposed their own import bans in turn.[4]

But if things were looking up in Asia, elsewhere the taint of the IT miracle lingered. Among the many regions beset by this twenty-first-century plague of desperate and perilous gleaning, Agbogbloshie in Ghana is infamous – a place where men, women and children labour side by side amid the poisonous stench of open fires to eke out an existence rendering down the products of luxury:

> … *pillars of black smoke begin to rise above the vast Agbogbloshie Market. I follow one plume toward its source, past lettuce and plantain vendors, past stalls of used tires, and through a clanging scrap market where hunched men bash on old alternators and engine blocks. Soon the muddy track is flanked by piles of old TVs, gutted computer cases, and smashed monitors heaped ten feet high. Beyond lies a field of fine ash speckled with glints of amber and green — the sharp broken bits of circuit boards. I can see now that the smoke issues not from one fire, but from many small blazes. Dozens of indistinct figures move among the acrid haze, some stirring flames with sticks, others carrying armfuls of brightly colored computer wire. Most are children.*[5]

According to the Blacksmith Institute, there are at least 500 of these digital hells worldwide, chiefly in Asia, Africa and Latin America, each of them the product of the casual disposal of last year's mobile phone, tablet or laptop by an affluent consumer.

In 2015 Achim Steiner, then Executive Director of the UN Environment Programme (UNEP), told journalists of a 'tsunami of e-waste rolling out over the world', warning that many elements found in electronic equipment are hazardous to people and the environment. 'Never mind that it is also an economic stupidity because we are throwing away an enormous amount of raw materials that are essentially re-useable,' said Mr Steiner.[6]

Some 500 different compounds, many of them toxic, are used in the manufacture of electronic goods. Typically, the e-waste stream contains epoxy resins, fibreglass, PCBs, PFOA, flame retardants, polyvinyl chloride, plastics, lead, tin, copper, gold, silicon, beryllium, carbon, iron, aluminium, cadmium, mercury, thallium, americium, antimony, arsenic, barium, bismuth, boron, cobalt, chromium, europium, gallium, germanium, gold, indium, lithium, manganese, mercury, nickel, niobium, palladium, platinum, rhodium, ruthenium, selenium, silver, tantalum, terbium, thorium, titanium, vanadium and yttrium. A single device, such as a smartphone, may have as many as forty of these substances in its makeup. Their health effects range from birth defects to nerve and brain damage, distorted blood composition, damaged lungs, livers and kidneys, and death by poisoning.[7] Many of these ingredients are rare – it may take a tonne or more of tailings to extract a single gram of precious metal or rare earth, meaning that the phone's pollution footprint can be millions of times larger than the device itself.

The output of the global electronics industry is colossal. Since the early 1990s, it has churned out more than 5 billion personal computers and 20 billion mobile phones, 10 billion televisions and radios, 1.5 billion tablets, music players, games platforms, lights, countless refrigerators and other whitegoods. Such is the spectacular rate of replacement of old electronic devices with new ones featuring the latest technology that many new devices

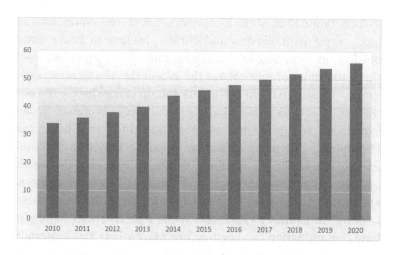

Figure 5.1. World growth in electronic waste in million tonnes, 2010–20. Source: UN University.

are rapidly superseded and become redundant – or at least 'unfashionable' – in a matter of months and are then discarded by their owners: the average life span of a desk-top computer has fallen from six to just two years. Many devices are designed to become obsolescent quickly in order to encourage consumers to buy upgraded, faster, more powerful machines to run ever-more demanding software. This practice redoubles the amount of e-waste in circulation around the Planet every few years (Figure 5.1). Customer behaviour in the electronics sector, more than any other, epitomises the grip of consumerism over the human psyche, and the mindless flood of toxic waste that results.

'50 million tonnes of e-waste are produced each year and, left unchecked, this could more than double to 120 million tonnes by 2050,' the World Economic Forum has warned. 'It is hard to imagine even 50 million tonnes, yet this is equivalent in weight to all the commercial aircraft we have ever built throughout history.'[8]

Professor Ming Hung Wong, from Hong Kong Baptist University, says the process of crude recycling of e-waste causes vast contamination of the local environment (soil, water, air and food crops) with dioxins and furans, leading to heavy body burdens in local people – children and infants especially – of up to 100 times or more above the World Health Organization's limits. 'These persistent pollutants end up everywhere – the air, the ocean, or leak into soil and groundwater,' Professor Wong explained. The problem has been identified in China, the Philippines, Vietnam, Pakistan, Mexico, Brazil, India and Africa. 'It's no longer confined to the villages that deal with e-waste, because when water and soil is polluted, everyone is vulnerable to the food products that are exported from these regions,' he cautions.[9] Professor Wong's observations highlight how contamination can begin in one place where raw materials are extracted, move to another where devices are made, move to a third far away where they are sold, used and discarded, sent to a fourth as scrap, reprocessed in ways that pollute air, soil, water and food and then even return to consumers on the other side of the world, either in exports of food grown in the polluted region or else via wind, water, fish and wildlife. It is a fresh example of the universal dissemination of chemical pollution in our interconnected world. Moreover, it shows that even citizens living in countries with sound regulation may be exposed to the consequences of their own heedless dumping, courtesy of the Anthropogenic Chemical Circulation (ACC).

Like climate change, e-waste is a toxic curse generated by the affluent, whose worst impacts fall disproportionately on the poor.[10] The scientific study of the health impacts of these post-modern 'dark satanic mills' is still in its early days, but researchers report higher incidences of oxidative stress, DNA damage and inflammation among the local people involved with e-waste or exposed to its emissions, the precursors of heart and lung disease and cancers.

One of the problems with e-waste is its extremely long toxic life span. Most of the e-waste produced since 1990 has been improperly disposed of and is largely still in circulation in the Earth system in air, water, soil, groundwater and the food chain. The cumulative exposure of humanity and all life on Earth to e-toxins is probably rising exponentially. The real health effects are yet to be seen, partly because of the difficulty of tying a particular health problem to a specific substance in the waste stream, and the chronic nature of many forms of poisoning.

In recent years, less than a quarter of the world's e-waste has been properly recycled – the remainder being dumped, either in poor countries or local landfills or inadequate waste facilities. This is in spite of the fact that 50 million tonnes of e-waste contain raw materials worth an estimated \$63 billion if recovered.[11] Even in Guiyu, the trade-off for sickness and appalling conditions was transient wealth.

E-waste offers a gruesome parallel with another industrial poison, asbestos, which has been used worldwide as a cheap building and insulation material, from the 1860s right up to the present day. The first cases of asbestos-related disease and death were recorded in 1906. Between the 1930s and the 1950s, knowledge of its dangers rose. However, it was so useful and profitable that Western governments and most of the asbestos industry long tried to keep the public in the dark about its deadly side-effects, so they could continue to mine and sell this valuable material. By the 1980s growing public health concerns led to bans on the mining and use of asbestos in over fifty developed countries and its progressive removal from buildings and homes; long and costly lawsuits began, as victims sought damages. However, even today when we understand its deadly nature, asbestos continues to be legally used in over a hundred countries: 125 million workers remain exposed to it and 107,000 die each year.[12]

E-waste differs from asbestos in one respect: the risks have long been canvassed worldwide in the media by concerned scientists and green groups, public concern is high and plans have been proposed to limit the harm it may be causing. UNEP, for example, has urged the adoption of centres of excellence in e-waste processing in countries all around the world. Since China banned it, most developed countries now make a greater effort to deal with their waste within their national borders rather than dumping it on the developing world. Progress is slow, however, and the cost of disassembly and metal extraction is high. Meanwhile, the flood of discarded devices swells with each passing week. In an industry where conspicuous consumption is driven by cheap retail prices, these complex devices are not designed for quick, easy and safe disassembly and recycling of their intricate components, and particularly not for the separation of their metals: to develop such designs would render them too expensive. Nevertheless, sixty-seven countries have so far passed laws requiring recycling or safe disposal of their e-waste, while Apple,[13] Google,[14] Samsung[15]and other prominent brands now accept 'trade-ins' of old phones on new ones and have begun setting targets for recycling and the use of renewable materials.

Although it remains far from cleaning up its act, the electronics industry has nevertheless begun moving slowly away from the 'take, make and discard' model that pollutes the Planet and towards becoming a pioneer of the circular economy that will sustain civilisation into the future.

Nanopollution

The advent of nanotechnology, the production of tiny particles measured in billionths of a metre, has – literally – added a new dimension to the universal pollution of people and the environment. Nanoparticles are so small they can enter the human body with ease, slip through the lungs into the bloodstream, pass through the skin, transit the blood–brain barrier into the brain, or breach the mother–embryo barrier to enter the unborn child. When particles are so small, they often have quite different physical and chemical properties from larger objects – which is why scientists use them. However, the same applies to their actions on the human body.

A familiar use of nanotechnology is the inclusion of tiny titanium dioxide particles in sunscreens and cosmetics, due to their superior ability to block or reflect the sun, so reducing the amount of skin damage and the 'ageing' it causes. Other uses include body armour, 'smell-proof' socks, sporting equipment, rain-proof jackets, electronic devices, paints and surface protectants, blood tests, bomb detectors, pesticides, energy devices, water filters, self-cleaning glass, antimicrobial bandages, skin-penetrating cosmetics, liquid crystal displays, bio-batteries and many more.

The number of nanoproducts released is soaring. One database listed 3840 consumer products available on the European market alone, covering mainly personal care, clothing, sporting goods and cleaning products.[16] Owing to industrial secrecy and the uncontrolled release of new products in many countries, this is undoubtedly a mere fraction of the actual number of nanoproducts entering the world market, as the USA and China are by far the largest generators of new research into these products. As with most outputs of the chemical industry, the genie was well and truly out of the bottle before anything was done to see whether the technology was safe for humans or the environment or not. Ever since, governments have been playing catchup – and losing.

There were two ominous warnings: quartz and asbestos. Tiny particles from both have been known to cause fatal disease in humans for over a century. Since then, the mass deaths of hundreds of millions of people from air pollution (also mainly inflicted by invisibly small particles) have rammed home the lesson.[17] With nanoproducts, it has largely gone unheeded.

In an early review of the safety issues around nanoscience, Canadian researcher Cristina Buzea and colleagues stated:

> ... *nanoparticles, have the ability to enter, translocate within, and damage living organisms. This ability results primarily from their small size, which allows them to penetrate physiological barriers and travel within the circulatory systems of a host ...*

> *A large body of research exists regarding nanoparticle toxicity, comprising epidemiological, animal, human, and cell culture studies. Compelling evidence that relates levels of particulate pollution to respiratory, cardiovascular disease, and mortality has shifted attention to particles with smaller and smaller sizes.*[18]

A recent review focusing on titanium dioxide (TiO_2) nanoparticles, which are widely used in sunscreens, cosmetics, paints and as food whiteners, concluded: 'Regular supply of TiO_2 NPs at small doses can affect the intestinal mucosa, the brain, the heart and other internal organs, which can lead to an increased risk of developing many diseases, tumours or progress of existing cancer processes. The mechanism behind the nanotoxicity of NPs has not been discovered yet. Many studies attribute it to oxidative stress.'[19]

Industrial nanoparticle materials today constitute a small but growing pollution source that is, so far, obscured behind mountains of other human emissions. Like earlier chemicals they are worming their way into the human food chain with hardly any oversight or control by governments – with few questions asked, until people start to sicken and die.

Not all nanoparticles are the direct result of industrial manufacture. In many cases they are released by the breakdown of other products, such as plastics and synthetic textiles, into ever smaller and smaller particles and fibres, as mentioned in Chapter 2. Nanoplastics are increasingly found in the human food supply, especially in wild-caught fish.[20] According to Brigitte Toussaint of the European Commission, 'the human health impact of micro- and nanoplastics contamination of our food and beverages remains largely unknown'.[21]

One of the scariest dimensions of nanopollution is that it may be a problem with no end to it: once released into the global environment, nanoparticles can never be retrieved. Being invisibly small, they are readily transported in the air, in water and in living creatures and, as we know from other substances, this pollution can rapidly disseminate Planet-wide as part of the ACC. No nation's regulation can protect its citizens. If certain particles prove dangerous and durable, they can potentially cause unlimited harm, forever – unless immobilised or broken down by natural processes. Even nanoparticles that are embedded in paints, metal objects or electronic goods may after a time be released into the environment, when those products degrade or are intentionally recycled: today's nano emissions may thus fall as a fresh curse on future generations for ages to come, unregulated, uncontrolled.

In a sense, every nanoparticle produced adds cumulatively to worldwide nanopollution and while this is still small relative to other sorts of contamination, it is growing at dramatic rates as millions of new products – like fleece jackets or rain-proof parkas – are made and released onto the market. The fact that many of these products are intended for personal use on the skin and in food increases the chances of a tragic outcome. There is a risk that the story of nanoproducts may thus echo that of tobacco smoking, which has led to a number of deadly diseases which were never suspected when tobacco was first introduced. Another parallel with cigarettes is that the term 'nano' has entered the vocabulary of the advertising industry to denote 'novel' and 'exciting', endowing such products with a fashion glamour calculated to ensure wide uptake and obscure the threat they pose.

Those risks grow with each passing day. When concerns about nanotechnology and its impact on human health and safety first surfaced in the early years of the twenty-first century, a number of international databases were established to track nanoproducts and to record all the available scientific evidence on their health and safety. With the passage of time those databases have ceased to exist – quite likely because, like climate science and tobacco science, they appeared to threaten the emerging industry and its profits.[22]

In 2017, in an article in *Nature*, Professor Steffen Foss Hansen from the Technical University of Denmark called for the mandatory reporting of all products containing nanomaterials.[23] However, outside of Europe such a prospect appears slim: globally there are no requirements for manufacturers to safety-test, register or arrange safe recycling of nanoproducts – so-called 'cradle-to-grave' product management – despite the emergence of medical groups such as SafeNano who advocate 'responsible nanotechnology development' and risk assessment.[24]

A major unknown risk with nanoparticles is their potential to interact with other poisons in our food, air, water and living environment and the diseases they cause. Just as heavy smokers were found to be at greater risk from the coronavirus,[25] so too those with conditions brought on by other forms of pollution or poisoning may find them exacerbated by nanosubstances. While questions such as these remained unanswered, the world nano-machine grinds on, pouring out its untested products in ever-increasing, heedless volumes.

Nanotechnology, for all its prospects and virtues, is a fresh reminder that we humans, and all life on Earth, are mere guinea pigs in a vast global chemistry experiment, for which nobody takes responsibility. The experiences of the past century, and the tens of millions of lives sacrificed to 'industrial progress', have taught us nothing.

Nutrient Nightmare

One of the largest of human polluting emissions, in terms of sheer volume as well as impact, is our release of nutrients into the world's waters. Yet this is also one of the effects least noticed by the public, as it usually involves the gradual transformation over years of clear, clean, healthy rivers, lakes and seas into turbid, stagnant and sometimes lifeless systems. These slow changes have been measured and recorded by science for over 150 years. However, in recent times, driven by a warmer climate, sudden changes have begun to appear, where a previously stable waterbody reached a tipping point and flipped into a stagnant, near-lifeless state.

The main culprits in this progressive destruction of the world's fresh and marine waterbodies are agriculture, transport and urban waste. Together these inject hundreds of millions of tonnes of nitrogen, phosphorus, potassium and other nutrients into the Earth system, in concentrations far higher than occur normally in nature. The process they unleash is known as eutrophication, or nutrient pollution (Figure 5.2). So far it has contaminated over 760 large areas of ocean or river estuaries.[26]

The main sources of these nutrients are agricultural fertil-isers, legume crops and animal manure, eroded soil, the burning of fossil fuels, washing detergents, urban runoff and the dumping of waste and raw sewage. These lost nutrients are

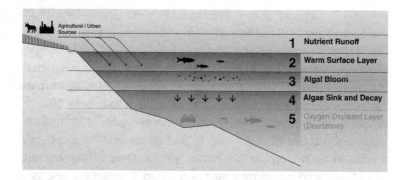

Figure 5.2. The eutrophication process. Source: Ocean Health Index, 2020.

causing profound changes in our rivers, lakes and oceans, turning once-productive waters brimful of life into empty, stagnant 'deserts' where few fish and other oxygen-reliant life forms can survive. Examples of nutrient pollution include:

- Contamination of ground and surface waters in India with untreated sewage, making them undrinkable and a major source of diseases such as 'blue baby syndrome'.[27]
- Declining oxygen levels in the world's oceans, reducing their ability to support life.[28]
- Formation of a 'dead zone' off the mouth of the Mississippi River, USA, over 15,000 square kilometres in extent.[29]
- Sediment and fertiliser runoff are responsible for killing a large part of the corals on Australia's Great Barrier Reef.[30]
- Pollution of China's rivers from farms and livestock has increased up to 45-fold and continues to rise.[31]
- Widespread nutrient pollution of Europe's Baltic Sea, likely to take centuries to repair.[32]
- The northern Mediterranean faces growing pressure from runoff, sewage, fertilisers, industry and aquaculture; stronger regulations are helping to curb it.[33]
- Florida's coasts are plagued by toxic 'red tides' driven by nutrient pollution.[34]
- Fouling of North America's Great Lakes so they become unsafe to drink.[35]

- The Philippines' famed Laguna de Bay, its largest lake, has been almost ruined by nutrient pollution and invasive plants.[36]

To give some idea of the scale of the challenge, the International Nitrogen Institute (INI) has announced a goal of reducing human emissions of nitrogen pollution by 100 million tonnes by 2030.[37] Recently it was estimated that humanity adds 120 million tonnes of reactive nitrogen and 20 million tonnes of phosphorus a year, cumulatively, to the Earth system. Over fifty years our use of nitrogen has grown sevenfold, and phosphorus use has tripled.[38]

Nutrient pollution is not new – the first cases of 'dead rivers' flowing through industrial towns date back to the 1850s. What is clear, and the reason for including it in a chapter about new and under-rated chemical threats, is that it is on track to double, and possibly triple, by the 2060s – as world food demand rises to meet the needs of over 10 billion people. Doubling food output from agriculture implies a probable doubling in the use of nitrogenous and phosphatic fertilisers, and a doubling in soil erosion, food waste and sewage entering the environment. This will have a potentially horrendous impact on all the world's main waterbodies, the life in them, and the ability of society to drink or use them.

Many waterbodies will suffer major tipping points. These in turn will eliminate entire fisheries, ruin tourism, prevent fish farming, threaten human health and accelerate dangerous climate change. For example, as the oceans stagnate, the microalgae and plankton which absorb carbon and produce oxygen are gradually replaced by anoxic bacteria which produce poisonous hydrogen sulphide gas and kill off other water species. As a result, the oceans absorb less carbon dioxide and the atmosphere more – and global warming speeds up. This process, scientists now believe, was partly to blame for the worst extinction event in Earth history, The Great Dying at the end of the Permian era, when 90 to 95 per cent of all life forms died out.

So nutrient pollution, while attracting less attention than other, more dramatic forms of contamination, nevertheless has potential to imperil the human future at a global scale. Indeed, the Stockholm Resilience Centre has warned, it already exceeds the safe boundary for human survival.[39] Despite some progress in reducing nutrient inflows to local waterbodies, there has been little or no advance in attempts to control worldwide nutrient pollution of the Earth system.

Fortunately, there exists a feasible technical solution: to close the nutrient cycle and re-use all our wasted nutrients in food production, over and over again. However, this calls for the extensive redesign of cities and food systems worldwide – as described in my book *Food or War*.[40] No nation in the world, however, is yet seriously contemplating such a profound shift and the global nutrient flood remains a serious blind spot if we are to live sustainably on this Planet.

Novel Pesticides

When it comes to contaminants, pesticides have drawn the main glare of public and regulatory attention, with the result that some – especially organochlorines and organophosphates – have been banned in well-run countries (though many developing countries still make them and, through the ACC, they still menace the Earth system and all its inhabitants). Despite these restrictions, world pesticide use has doubled since the early 1990s, rising from 2.2 million tonnes to 4.1 mt in 2017[41] (Figure 5.3) and reaching 5 mt in 2022: this is about tenfold larger in sheer tonnage than it was when Rachel Carson first warned the world about its impact, in the early 1960s. These numbers are only intended to indicate the growth in demand, which was rising by about 22,000 tonnes per year – and convey no estimate of overall toxicity. Indeed, toxicity may be rising far faster than tonnage, as older, less efficacious pesticides are replaced by newer chemicals that are more potent in lower doses, though generally more specific in action.

It is essential to note that, without the use of pesticides (herbicides, insecticides, fungicides, nematicides and other crop protectants) the world crop harvest would be from a third to a

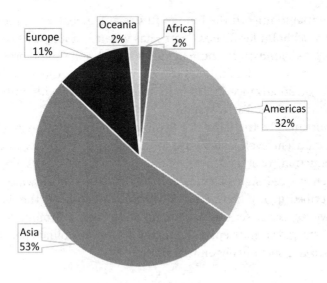

Figure 5.3. World pesticide use, 2017: 4.1 million tonnes. Source: FAO.

half smaller,[42] and this would cause the number of malnour-
ished and starving people to blow out from 0.8 billion to over 3
billion, with particularly dire effects for the world's most populous countries. Pesticides therefore play a key role in protecting
the agricultural food supply: the issue is whether they need to do
it at the expense of human and ecological health and safety, or
whether there are safer, less toxic means available. Also,
whether food systems in developed countries require such high
chemical intensities.

World food demand is forecast to rise by 70–75 per cent by
2050 and to double by the early 2060s. Most of the increase is
expected to come from higher crop yields and, climate permitting, this implies a doubling in the main agents of high yield –
fertilisers and pesticides. Chemical pesticide use will thus reach
10 million tonnes a year by the 2060s, unless there is radical
change in global food production systems. Total emissions
between now and then could be 280 million tonnes of specialised
poisons, representing a vast shock to the biological health of the
Earth system. This is not an issue that can be lightly dismissed.

However, this upsurge in pesticide use will also coincide and combine with the even vaster impact of other forms of chemical pollution, such as e-waste, hazardous waste, mineral waste, fossil fuel pollution, plastics and so on. The tendency of governments, industry and consumers has been to view these as isolated problems, rather than as components of an aggregate toxic shock to the Earth and all its people. If such blinkered attitudes persist, then the scale of this shock will be under-rated rather than over-rated, under-reported and under-researched. It will be far worse. Many pesticides are already linked in the scientific literature with leukaemia and other cancers, with birth defects, infertility, brain, nerve, skin and developmental disorders, respiratory diseases, diabetes and depression – and their interaction with many other sources of poison has grave implications for the health of the human species and for life on Earth.

New pesticides are often promoted by the chemical industry as 'softer', with effects more specific to the target pest, less toxic to humans and kinder to the environment and, indeed, many are designed with this in mind. Whether this is in fact true remains open to scientific question as there is a lack of verifiable, independent evidence.

An example in point is the global use of neonicatinoid insecticides,[43] a class of insect nerve poisons known collectively as 'neonics', introduced in 1990 for crop protection largely because they were thought (or claimed) to do less damage to birds and animals than the organophosphates they replaced. The neonics quickly became the world's most widely used insecticides. What has since emerged echoes the blunder made when ozone-layer-destroying CFCs were replaced with climate-destroying HCFCs as spray can propellants. Since the early 2010s there has been a blizzard of scientific papers reporting disastrous consequences for honeybees and other pollinators, for insects generally – and hence for the birds, reptiles, small mammals and fish that rely on insects as food.[44]

Around three-quarters of the world's food crops rely on insects to spread their pollen, so they can produce fruit or grain. Over 35 per cent of the world's croplands, and eighty-seven different crops, require the pollination services of insects to yield a crop.[45] However, numbers of bees and pollinating insects have fallen to crisis levels in recent years, and most of the scientific research points at neonics as a primary culprit.[46] 'The decline across insect orders on land is jaw dropping,' Michigan State University entomologist Nick Haddad told *Time* magazine. 'Ongoing decline on land at this rate will be catastrophic for ecological systems and for humans. Insects are pollinators, natural enemies of pests, decomposers and besides that, are critical to the functioning of all Earth's ecosystems.'[47]

Europe reacted with alarm to the wipeout: in 2018, after close scientific study, the EU announced a comprehensive ban on the main neonics.[48] The USA, on the other hand, announced it would continue to use them, subject to minor restrictions.[49] It was an echo of what had already happened with pesticides, food additives and cosmetics, with Europe taking a precautionary stance, based on the amassing scientific evidence, while the USA sided with the manufacturers.

The scope for harm to insects worldwide and the pollination services they provide to humanity – valued at an estimated half a trillion dollars a year – is a case of the two-edged impact of man-made chemicals. On the one hand, the neonics lift food output by curbing insect damage. On the other hand, they reduce it by killing pollinators. Meantime there is mounting evidence of adverse effects, both acute and chronic, on human health.[50]

A central figure in the row, Professor David Goulson of Stirling University, commented:

There are obvious parallels with the tobacco industry. For 50 years they insisted that smoking wasn't harmful to human health, even when the scientific evidence piled up, they still claimed there was no link. And they funded scientists to come up with spurious studies, which seemed to back them up. Here, the scientists funded by the [agrochemical] industry are the only ones that stand up and say, as far as we can tell, there's no effect of these pesticides on bee health. I can't help but being [sic] highly cynical about the independence of any of that [research].[51]

Goulson's argument raises a chilling question: if every new chemical and pesticide were to undergo half a century of prevarication by manufacturers before its safety or harmfulness can be established, it will be centuries before all the dangerous ones are investigated and eliminated. What will be the human and Planetary cost of such a delay?

Chemical Intensification

As more scientific studies show, today's children first encounter petrochemicals in the womb, and then in their mother's milk. Seemingly unaware, parents then clothe their children in garments made from petrochemicals, feed them on food containing petrochemicals, give them toys made from petrochemicals to play with, bathe them in petrochemicals, put them to bed in them and surround them in their daily lives with homes furnished and decorated and cars made from the byproducts of oil. Then they wonder why their children are becoming increasingly sickly and suffering more unexplained conditions. So they take them to the doctor, who prescribes medication which is usually made from petrochemicals.

A pivotal issue for the human future is the question of 'chemical intensification', which is the widespread replacement of natural materials with outputs of the oil industry in industrial and commercial products such as lubricants, coatings, adhesives, inks, dyes, creams, gels, soaps, detergents, fragrances, textiles and furniture. Many objects in daily life that used to be made of wood, natural fibres, vegetable oils, clay, metal or glass are now made from oil-based plastics or synthetic fibres: furniture, tableware, toys, ornaments, clothing and footwear, bedding, shopping bags, packaging, tools, building materials, vehicle and aircraft parts. In the space of a generation society has gone from being largely natural product-based to largely chemicalised. Even 'wood' used in furniture nowadays consists, more often than not, of wood fibres artificially bound together with epoxy resins: so-called particle board.

In 2012 UNEP uttered a public warning about the dangers of this universal trend.[52] 'Chemical intensification is not just a measure of the chemical production and use but reflects changes in functions of chemicals and the importance of chemicals in all aspects of economic development. It also incorporates the increased complexity of chemicals themselves and the ever lengthening and more intricate chemical supply chain,' it explained.[53]

This abundance in diversity and use was likely to propel the expansion of worldwide chemical production at a rate of about 3 per cent a year between now and 2050, UNEP anticipated – much faster than the rate of growth of the world economy, pointing to a doubling, possibly a tripling, in worldwide petrochemical use and release into the environment by 2050.

In 2002, the World Summit on Sustainable Development (WSSD) held in Johannesburg, South Africa, declared a goal 'to achieve, by 2020, usage and production of chemicals in ways that lead to the minimization of significant adverse effects on human health and the environment'. By 2019, despite constant reminders and pleas for action in global fora, UNEP admitted, 'findings ... indicate that the sound management of chemicals and waste will not be achieved by 2020'.

'While many chemicals are important for sustainable development, trends are a cause for major concerns, requiring urgent action,' UNEP warned, once again.[54]

It found:

- 'The size of the global chemical industry exceeded $5 trillion in 2017. It is projected to double by 2030. Consumption and production are rapidly increasing in emerging economies. Global supply chains, and the trade of chemicals and products, are becoming increasingly complex.'

- 'Driven by global megatrends, growth in chemical-intensive industry sectors (e.g. construction, agriculture, electronics) creates risks, but also opportunities to advance sustainable consumption, production and product innovation.'

- 'Hazardous chemicals and other pollutants (e.g. plastic waste and pharmaceutical pollutants) continue to be released in

large quantities. They are ubiquitous in humans and
the environment.'

- 'Trends data suggest that the doubling of the global
chemicals market between 2017 and 2030 will increase
global chemical releases, exposures, concentrations and
adverse health and environmental impacts unless the sound
management of chemicals and waste is achieved worldwide.
Business as usual is therefore not an option.'

'UNEP calls for urgent action to tackle chemical pollution,'
reported Xinhua news agency in China, one of the world's most
polluted countries.[55] Despite far-sighted plans to try to lower the
chemical burden, such as SAICM (Strategic Approach to
International Chemicals Management) it fell pretty much on
deaf ears. A majority of the world's countries cold-shouldered
chemical law reform and only a handful of major corporations
signed up. The burgeoning chemical industry had won.

Megakillers

Despite negligible progress in containing ordinary chemicals,
not supposed to kill people, there had been heartening movement towards worldwide elimination of tailor-made mass
killers, under the UN Convention on Chemical Weapons. By
2020, it reported, 70,000 tonnes (97%) of the world's known
stocks of 72,000 tonnes of chemical weapons had been verifiably
destroyed; 194 countries had signed the convention and only
three stood outside it; seventy-four of the world's chemical
weapons plants had been destroyed and the remainder converted to peaceful purposes.[56] It had all taken a little under
twenty-three years. On the face of it, what had been achieved
with chemical weapons could potentially be achieved with
nuclear arms.

However, a lurking concern remained over the fate of tens of
thousands of tonnes of chemical munitions dumped into the
world's oceans between the end of World War 1 and 1972, when
a convention banning ocean dumping was declared. Nobody
knows the exact quantities, locations or types of the dumped

chemical agents: these facts are lost to history and so render any serious clean-up effort impossible. The US Army alone is said to have 'dumped 64 million pounds [29,000 tonnes] of nerve and mustard agents into the sea, along with 400,000 chemical-filled bombs, land mines and rockets and more than 500 tons of radioactive waste' at twenty-six sites that ring the US coastline.[57] A report to the Congress in 2007 stated that there were thirty-two American dump sites off the US coast, and forty-two off the coasts of other countries, notably Japan.[58] Around 500 people, mostly fishermen, have been injured by ocean-dumped chemical munitions in the last half century, but the real concern is over what happens when the containers of these ultra-toxins begin to break down and they leak into the ocean, the global food chain and the Anthropogenic Chemical Circulation. Most nerve agents tend to dissolve in water, but substances such as sulphur mustard can remain dangerous for years. It is a classic example of how yesterday's chemistry returns to haunt humanity living in the future.

Criminal Chemicals

The production of chemicals is so profitable that it has become a financial mainstay of organised crime around the world, worth half a trillion dollars a year.[59] Illegal recreational drugs are one of the fastest growing, uncontrolled sources of toxicity on the Planet. While their users generally regard them as fairly harmless – or, at worst, risky only to themselves – there is in fact a swelling tide of toxic contamination of soil, groundwater and urban wastewater systems and even the food chain, with potential to harm hundreds of millions of people who don't take drugs. This applies especially to amphetamine-family drugs, synthetic cannabinoids and their chemical precursors – which start in the legal pharmaceutical sector and are then 'cooked up' into street drugs by backyard operators who know little about chemistry and even less about safe production and waste disposal. Despite a brief downturn due to the coronavirus pandemic, the market for illicit drugs is thought to have expanded by around 50 per cent over the 2010s, with about 300 million users worldwide, of whom about 35 million suffer drug-related diseases and 12 million die each year.[60] This stark number is in addition to the other 9 million chemical-related deaths described previously.

The main routes for toxic substances from illegal drug use into the wider environment are in the urine of users and via the improper disposal of chemicals during manufacture, down drains and toilets. Homes and trailers used for 'cooking' illegal drugs are permanently contaminated with carcinogens and other toxins; these sites are deemed uninhabitable by local authorities and are often destroyed.

Recreational drugs are popular because a very tiny quantity can produce quite strong and lasting effects on the brain and nervous system of the user – and it is this aspect that makes them such a dangerous part of the ACC. Many are endocrine disruptors, meaning that tiny amounts can also interfere with the body's hormonal and reproductive systems, and there is growing evidence that they may cause cancers. Recreational drugs are also strongly linked to acute effects, such as heart attacks and poisoning, and are associated with chronic effects such as lung disease and psychosis. Regrettably, the present focus of concern is far more on the health of drug users than it is on the health of communities exposed to their noxious emissions.

Use of recreational drugs generally starts with young people out for a thrill. At this early point, work in tobacco and alcohol shows, individuals are open to educational messages about the risks of self-harm. However, an important additional message – as yet unheard – is that by taking drugs they contribute to the poisoning of society, wildlife and the local and global environment. People who poison water supplies face criminal charges – but for drug users, who also poison society's water, there is no such penalty. There is no easy solution to the global illicit drug problem, but awareness that drug use is now an important contributor to the poisoning of our Planet – and thus a practice inimical to the future of all humans and all life – may help deter some young users.

Seeing the Unseen

The categories of contamination described in this chapter share one thing in common: they are rarely considered poisons by the public at large, and their combined potential to injure our health and wellbeing into the future is very poorly understood.

This underlines the urgency for educating a new breed of consumer: not one who witlessly seizes the first enticing product in the supermarket or electronics store, but one who inquires not only into the product and what it contains but also into its past and its future: the processes by which it was made and how it may be disposed of.

Without such consumer sentiment steering manufacturers towards safer products, humanity is fated to poison itself in ever-increasing doses for the remainder of its history.

Without such consumers, ethical industries will lack the economic incentives they need to introduce clean, safe, sustainable production methods and products. Without such consumers, 'cheap and dirty' industries will continue to put the clean guys out of business.

The dirty guys will win – and all humanity will lose.

In Chapter 9 we will explore ways such an enlightenment might come about, but for the time being it is enough to be mindful that our future is in our own hands. What that means for our health is explained in the next chapter.

6 SICK SOCIETY

'By drilling deep into the bowels of the Earth ... we have unwittingly and catastrophically altered the chemistry of the biosphere and the human womb.'
Dr Theo Colborn: Letter to the President, TEDx, 2012

In the first ten years of the twenty-first century, more than 3000 children were poisoned in the Australian town of Port Pirie, with the full knowledge of governments, health officials, regulators, industry and the community itself.[1] The poisoning had actually been going on for more than 120 years, ever since the first lead smelter was built in the small, regional South Australian industrial centre; the first public warning was sounded by a Royal Commission inquiry as far back as 1925, with reiterations being made on numerous occasions, particularly in the 1980s and 1990s. In 2012 a further investigation found lead levels presenting serious risk of harm in the children's playground.

'The adverse impacts associated with (lead) production have been consistently downplayed by industry, governments, councils, health officials and regulators,' said Professor Mark Taylor of Macquarie University, who investigated the case. 'Even some academics argue the effects of low lead exposures are not of significant concern. Due to ignorance, misinformation, and deliberate obfuscation of evidence, generations of families living next to lead-mining, smelting, and refining centres such as those in Broken Hill, Port Pirie and Mount Isa, have been and continue to be exposed to environmental lead, a known neuro-toxic contaminant.'

His statement was based on a series of ground-breaking studies by scientists over several decades. In just one of these, a 1992 investigation involving 492 children living around the Port Pirie smelter, researchers established a clear correlation between raised lead levels in the children's blood and reduced IQ: children with up to 30 micrograms of lead per decilitre of blood were found to have lost between 4 and 5 per cent of their intelligence.[2]

The lead dust lay like a drear, grey mantle over the homes and suburbs surrounding and downwind of the smelter. Samples taken two decades later, in 2011, by Taylor and his team from the children's playground equipment returned average lead values 173 times higher than those at another playground in a nearby town, Port Augusta, which had no smelter. Not content with sampling only the equipment, the researchers took samples from children's hands after they had been playing outdoors for a period of time: 'Hand lead is of greater concern because young children tend to put their hands in their mouths, and it is therefore a significant pathway for childhood lead exposure,' Taylor explained. After just twenty minutes' use of the playground equipment, children typically showed an average daily dose more than forty times above recognised safety limits. Although born with relatively low lead levels in their blood, by two years of age the children were already showing lead contamination above internationally recognised levels of concern – mostly inhaled or ingested in their own homes. The exposure continued as they grew older and began to attend pre-school and junior school.

'Childhood exposure to lead has been linked to lower IQ and academic achievement, and to a range of socio-behavioural problems such as attention deficit hyperactivity disorder (ADHD), learning difficulties, oppositional/conduct disorders, and delinquency. The disabling mental health issues from lead exposure often persist into adolescence and adulthood,' Taylor explained.

'There has been and continues to be significant government knowledge of the true nature and extent of the problem. The fact that the politicians have not acted on the evidence demonstrates they have ignored the information coming from their staff, and/or have not had the willpower or commitment over the last 30 years to take effective action to eliminate preventable and damaging exposures once and for all.'

The authorities responded to the threat to child health and safety by advising the parents to wash the children's hands more thoroughly and to clean their homes more thoroughly – a recommendation made primarily because the smelter was continuing to belch 44,000 tons of nerve poisons over the suburbs every year. Meanwhile, because of its economic importance to the community, little was done to curb the source. 'The only conclusion one can draw from the failure to eliminate preventable lead exposure in Port Pirie is that there has been an absence of decisive and competent leadership from successive governments over the last 30 years,' Taylor ended.

The poisoned children of Port Pirie are far from alone: worldwide, hundreds of millions of children are exposed daily to a battery of brain-damaging poisons emitted by a myriad sources – smelters, coal-fired power stations, toxic garbage heaps, traffic fumes, belching factories and contaminated food and drink – and suffer lifelong impairment to their precious, developing intelligences as well as many other disabilities and conditions described in this chapter. A million children, mostly younger than five years of age, die from such causes every year.[3] In country after country, governments turn a blind, or at least indifferent, eye to the health and safety of their children, at the behest of industries which pour out chemical pollution and which, every year, pile on the pressure for repeal of what few legal protections humanity still has. In the USA, reported the *New York Times*, the Trump Administration rolled back over 125 environmental laws between 2016 and 2020.[4] Around the world, right wing governments, inspired by the USA's contempt for its citizens' lives and wellbeing, make similar choices, at the dictate of chemical corporates and their shareholders.[5]

In the case of familiar toxins such as lead, mercury, arsenic, asbestos and cadmium, governments are well aware this contamination is happening – or at least have very strong grounds for suspecting it – but prefer not to act because to stop poisoning children would be harmful to business profits, and nowadays business profits purchase political influence and success.

Babies and children, be it noted, have neither votes nor any voice in the matter. It is a universal human obscenity that they must suffer and die to feed the lust for wealth and power of a handful of heedless scoundrels in politics, heavy industry, science and the stock exchange.

A Dumber Race?

A tempting explanation for the widespread indifference of modern society to the mass poisoning of its children is that we are less intelligent than previous generations of humans.

After a marked increase in human intelligence during the first three-quarters of the twentieth century – a phenomenon known as the Flynn effect after the American psychologist who discovered it[6] – recent scientific evidence points to a clear downturn in human brainpower since 1975. The average rate of decline has been around three IQ points a decade, amounting to the loss of about 13.5 per cent in average intelligence between 1975 and 2020.[7] Similar results obtained in studies carried out in Norway, Denmark, the UK, France, the Netherlands, Finland and Estonia put the decline in intelligence beyond question, though rates vary. Beyond noting that the drop in intelligence must be caused by 'environmental factors', not by genetics, researchers have so far been unable to confidently assign specific causes. But then it took over fifty years before they were able to confidently connect lung cancer with tobacco smoke.

Rising exposure to lead, mercury, organochlorine pesticides and other substances known to inflict permanent neurological damage on the brains of the very young, reducing their intelligence lifelong, cannot be ruled out. Although our egos might howl with indignation at the mere suggestion that, on average, we're dumber than our parents, and although the average loss of intelligence per person is fairly small, such a possibility nevertheless highlights the potentially devastating intergenerational impact of exposure to a neurotoxic soup. More significantly, it foreshadows **a civilisation becoming progressively less intelligent with each passing decade**, to the point where it is no longer smart enough to take action in the face of mounting threats to its wellbeing, health, social stability and long-term survival.

Aside from exculpating our genes as the primary cause of the dumbing down of society, scientists remain uncertain what are the main drivers. Speculation includes our recent media habits, declining educational quality, and poorer diets and health. The application of William of Occam's Razor suggests we should probably investigate the best-evidenced theory first: the tens of millions of tonnes of toxic chemicals and nerve poisons released into the human living environment since the mid-1970s. The value of worldwide chemical sales has risen 46-fold over the past fifty years. It is forecast to reach $18 trillion in 2050. The volume of chemical production, already 2.3 billion tonnes a year, is forecast to grow by up to 450 per cent.[8] On average, human chemical use – and with it, exposure – has increased steadily, by around 4 per cent a year during the last half century, a rate far higher than either economic or population growth.

That some chemicals cause brain damage is not news. The first description of lead as a brain poison was written nearly two thousand years ago by the Greek physician Dioscerides (AD 40–90), who observed that 'Lead makes the mind give way'. In the seventeenth and eighteenth centuries, wine was often adulterated with lead salts, to the misfortune of its drinkers. In the twentieth century, tetraethyl lead was widely used as a performance additive in motor fuels, lead solder in food cans and lead in paints: together they poisoned millions. In 1892, children in Brisbane, Australia, originally diagnosed with meningitis, were found to be suffering from lead poisoning from paint. So there have been plenty of warnings.[9]

Scientific medicine has known for more than half a century that exposure to lead, mercury and organochlorines damages the brain and central nervous system of foetuses, babies and small children, and reduces their IQ, often lifelong. These findings were highlighted in investigations of the horrendous events at Port Pirie, Minamata and Flint, USA (where the ability of third-grade children to read dropped from 42 to 11 per cent in just three years).[10] Since then, they have recurred around the world. In 2009, China forcibly evacuated 15,000 residents from the fallout zone of a smelter in Jiyuan city, Henan province.[11] We know from archaeological study of bones and teeth that people today are far more heavily contaminated with these metals and toxins than they were in the past. In more recent times, with the progressive removal of lead from petrol, cans and paint in many countries, the incidence of brain damage from lead has fallen – but it has increased as a result of exposure to other substances such as flame retardants, pesticides and endocrine disruptors.[12] This is a stark illustration of the never-ending game of catch-up between poison-emitting industries and regulators – as soon as one poison is controlled, more take its place.

A loss of ten or fifteen IQ points may not sound all that significant to many people. However, even smaller reductions in intelligence are linked in the scientific literature to increases in murder rates, violent crimes, juvenile delinquency, poor school performance, low job performance, short attention spans and unwanted pregnancies in heavily affected populations. Low IQ is particularly observed in prison populations and appears linked to recidivism.[13]

These studies also underline one of the more insidious aspects of chemotoxicity: while it may not kill the person most affected, chemically induced brain damage may be an influential factor in them killing, raping or damaging someone else. The risks of such subtle yet life-threatening effects are rarely considered – and impossible to quantify – when new chemicals are released into society. However, they cause casualties as surely as do other directly toxic substances.

And, while loss of a few IQ points may not prevent an individual leading a relatively normal life, for populations at large the risks are likely to be cumulative, argued Harvard Medical School's Professor David Bellinger. He calculated that

American children between them had shed 23 million IQ points due to lead exposure, another 17 million due to pesticides and 300,000 due to mercury,[14] making America overall a less intelligent country. Other scientists concluded that chemical poisoning from these three substances causes more damage to the intelligence of society than premature births, autism, ADHD and traumatic brain injuries combined, and deserves to be rated alongside the other major diseases that afflict modern society.[15]

A profoundly disturbing dimension of the global loss in intelligence is that populations may be more easily influenced politically to vote against their own best interests, using simplistic propaganda tools on social media, such as those employed in elections in the USA, UK, Australia and Brazil.[16] Lacking the intellectual capacity to analyse what is told them, a growing segment of the population is ready to believe even the most baseless claims and outlandish conspiracies, is easily targeted by manipulative psychological warfare techniques – and induced to vote accordingly. Hard science for this hypothesis is still accumulating, but when the University of Montreal's Chenjie Xia and colleagues tested brain-damaged against normal patients in an election simulation, they found damage to the orbitofrontal cortex affected political judgement (as it has been found to do with economic and other forms of judgement).[17]

While it is still theoretical, the idea that population-wide brain damage due to toxic chemistry underlies the widely observed decline in Western democracy,[18] and the rise of populist or despotic governments, remains a dangerous possibility for civilisation: 'Democracy is undergoing an "alarming" decline across the world as a growing number of countries move towards authoritarian rule, according to the Freedom House think tank,' Britain's the *Independent* reported in 2019.[19]

Three substances – mercury, lead and organochlorine pesticides – have been singled out because they have the best-documented case histories for causing developmental disabilities and lifelong mental deficiency. They also especially affect the developing brains of infants and children while causing limited harm to adults. However, in 2006 Harvard's Dr

Philippe Grandjean warned: 'Two hundred additional chemicals are known to cause clinical neurotoxicity in adults. Despite a lack of systematic testing, many more are known to be neuro-toxic in laboratory models. Their toxicity to the developing human brain is not known and they are not regulated to protect children.'[20] By 2020 the number of substances confirmed as neurotoxic in humans or animals (causing brain and nerve damage) had expanded to 'nearly 1000', according to lawyers specialising in brain injury.[21]

In all likelihood, the number of chemicals that damage the human brain, mind and intellect runs into many thousands – but too little is still known about how they do it for governments to control their production and use, and industry is disinclined to try. The gaps in our knowledge of the harm caused by these substances, the way they interact with one another, plus the high levels of proof required to satisfy regulatory bodies that they should be banned, together make the task of protecting our children, their precious intelligence and our society as a whole, daunting – if not impossible – for the time being.

It is not only industrial chemicals and fossil fuels that can harm a child's intelligence. The modern food system is also a culprit, a study by Bristol University's Kate Northstone and colleagues found. 'There is evidence that a poor diet associated with high fat, sugar and processed food content in early child-hood may be associated with small reductions in IQ in later childhood, while a healthy diet, associated with high intakes of nutrient rich foods described at about the time of IQ assessment may be associated with small increases in IQ.' In their study, children nourished on healthy foods were almost three IQ points brighter than those fed mainly on processed foods.[22] Of course, processed foods also generally contain more poisonous chemicals.

In a society that professes to love its children and aspires to give them the best possible start in life, the harm now being inflicted on their developing minds is out of all proportion compared with only a few decades ago. Nowadays, said Grandjean, 'One out of every six children has a developmental disability, and in most cases these disabilities involve the nervous system.'[23] Evocatively, he branded the phenomenon 'a silent pandemic'.

Such statements raise disturbing questions: why do we strive to develop excellent education systems, while at the same time we damage the minds that are to receive them? How can we claim to care about our children when, year after year, we increase both the number and volume of the very things that cause them lifelong harm? Are rising crime, antisocial behaviour and terrorism rates attributable – even in part – to our failure to protect young minds from chemical damage? And especially, how long can we pretend to ourselves that we are not complicit in this damage, and avoid the fact that the substances which inflict it are the result of our own unbridled demand for consumer goods and an ever-higher material 'living standard', and the ungoverned efforts of industry and the market to satisfy us?

Children growing up in low-income societies suffer particular developmental and intellectual disabilities caused by their exposure to badly managed waste dumps, polluted air, water and food. A study carried out in India, Indonesia and the Philippines found levels of lead in the blood of children living near garbage dumps had cost each child, on average, six to eight IQ points, and increased the number of mentally retarded children in the vicinity by around 6 per cent, throwing a further, underappreciated, burden on the developing world.[24]

The implications of all this are profound. The loss of two or three IQ points per person per decade may not add up to a major difference in the short term. But the loss of seven or eight points in each successive generation is horrifying. It carries the inference that global society in the mid-twenty-first century could have a 28 per cent lower IQ than, say, the World War 2 generation. This would have, among other things, catastrophic consequences for the world economy, national economies, industry and employment in terms of lowered productivity: economic research has proven strong links between intelligence and national output.[25] It also implies a far larger proportion of the population unable to hold a job, leading to an explosion in welfare.

And, since low IQ is also quite strongly correlated with criminality – the typical criminal is thought to have an IQ of 92 (eight points below the norm) – a society-wide drop in IQ risks an upsurge in murder, violence, imprisonment rates, mental hospital intakes, gang warfare, illicit drugs and international crime as well as welfare outlays. That such an outcome might stem from increasing chemical damage to the brains of today's children is not a possibility that has been widely considered: society and science have, for the most part, yet to join the dots.

They are fast losing the ability to do so.

Chemical Disruption

In June 2009, the Endocrine Society, alarmed at burgeoning scientific evidence pointing to widespread chemical damage to the human hormonal (endocrine) system caused by certain disruptive substances or mixtures, released a powerful statement warning that 'even infinitesimally low levels of exposure— indeed, any level of exposure at all—may cause endocrine or reproductive abnormalities, particularly if exposure occurs during a critical developmental window. Surprisingly, low doses may even exert more potent effects than higher doses.'[26] It was the first clear, science-based departure from the old saw that 'the dose makes the poison', behind which the chemical industry had for so long taken refuge to justify its inaction over damage to the health and safety of millions.

The Endocrine Society is a worldwide body of medical professionals representing some 24,000 doctors in 120 countries. It speaks with authority and conviction, backed by reams of scientific evidence. In 2018, dissatisfied with the global response to its earlier warning, and in the light of a further 'exponential increase' in scientific research describing hormonal damage, it issued an even more powerfully worded statement, known as EDC-2.[27]

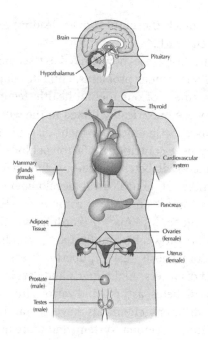

Brain
Pituitary
Hypothalamus
Thyroid
Mammary glands (female)
Cardiovascular system
Pancreas
Adipose Tissue
Ovaries (female)
Uterus (female)
Prostate (male)
Testes (male)

Figure 6.1. Endocrine-disrupting chemicals can harm any of our hormonal systems. Source: Endocrine Society.

Endocrine-disrupting chemicals (EDCs for short) are any chemical or mixture that can interfere with any part of the human (or animal) hormonal system, illustrated in Figure 6.1.

The human body has around fifty different hormones, key molecules that initiate a chain of command and control which governs things such as how we store and use our energy, grow, mature, reproduce, heal injuries, balance our blood pressure and blood sugar levels, produce milk, sleep and wake, regulate our brain and central nervous system, cope with stress. Familiar examples include insulin, thyroid hormone, growth hormone, oestrogen, androgen and testosterone.

These substances are critical to our survival: imagine the chaos that would ensue if, for example, you replaced one red city traffic light with a green one at a major intersection. By imitating our hormones chemically, EDCs can send our body false signals or block the natural ones, leading to dysfunction, disease and even death.

In particular, EDCs are now linked by science to cancers, obesity, sexual and gender problems, child developmental problems and impairment of our mental health, feelings and nerve responses. They are inextricably linked to the worldwide rise in 'lifestyle diseases'. 'Chemicals which disrupt the hormone system ... may be a contributing factor behind the significant increases in cancers, diabetes and obesity, falling fertility, and an increased number of neurological development problems in both humans and animals,' comments the European Environment Agency (EEA).[28] According to the Endocrine Society, EDCs are now also implicated in conditions such as infertility, endometriosis, early puberty, breast and prostate cancer, thyroid disorders, Alzheimer's and Parkinson's diseases, ADHD, obesity, diabetes, asthma and various immune disorders.

Some 1400 different chemicals have so far been found to disrupt the human hormonal system, and more are being added to the list each year.[29] They include synthetic oestrogen and well-known toxins such as PCBs, phthalates, parabens, PBDEs, PFOA, BPA and DDT. These create a body burden that may accumulate for as long as personal exposure continues – but which will usually decline when exposure ceases. These substances are to be found in such harmless-seeming items as cosmetics and 'personal care' products, plastic drink bottles, packaging and food containers, household cleaning products, sunscreens, shower curtains, soft toys, plastic cutlery, TVs, computers and electronic devices, carpets, bedding, wood treatments, medical equipment and medicines, tin cans, detergents, and food and drinking water.[30] They are also found in common medical drugs including contraceptives and in illegal drugs such as methamphetamines. As a result, these chemicals are now widely disseminated in fresh water and even the food chain, after being flushed down the toilet in the urine of users and then recycled in urban water supplies.

In the UK, for example, researchers found traces of the contraceptive hormone estradiol in 80 per cent of the rivers and lakes they tested.[31] Similarly, the US National Geological Survey found female hormone in 80 per cent of US drinking water sources.[32] This flood of female hormones is certainly feminising fish that live in these waters,[33] and may also be feminising men who drink the water, leading to observed phenomena such as shrinking genitals, swelling breasts, reduced sperm counts and sexual dysfunction. Around two men in every five now have fertility problems.[34] By 2045 most couples may require fertility treatment.[35] Oestrogens may sound fairly harmless, but for those affected the effects may manifest in a host of symptoms such as low sex drive, weight gain, muscle loss, migraines, depression, fatigue, mood swings, skin diseases, allergies, memory loss, hair loss, headaches and insomnia.[36]

The US National Institute of Environmental Health Sciences (NIEHS) makes the point that EDCs are now all around us and that exposure is therefore unavoidable. While a handful of EDCs have been banned in a few countries, the overall level of contamination continues to rise sharply throughout the Earth system and across the whole of humanity. Your system is clogged with these chemicals at this very moment, messing around with your bodyfat, your gender, your mind, your mood, your sexual preferences and your ability to have children.

Warnings about EDCs have been growing louder for three decades – but are still widely ignored by industry and most governments. The late Dr Theo Colborn, EDC pioneer and author of the trailblazing book *Our Stolen Future*, related:

> In 1991, an international group of experts stated, with confidence, that 'Unless the environmental load of synthetic hormone disruptors is abated and controlled, large scale dysfunction at the population level is possible.' They could not perceive that within only ten years, a pandemic of endocrine-driven disorders would begin to emerge and increase rapidly across the northern hemisphere. Today, less than two decades later, hardly a family has not been touched by Attention Deficit Hyperactivity Disorder, autism, intelligence and behavioral problems, diabetes, obesity, childhood, pubertal and adult cancers, abnormal genitalia, infertility, Parkinson's or Alzheimer's Diseases.[37]

The World Health Organization (WHO) and UN Environment Programme consider far too little is being done to abate the global deluge of hormone-disrupting chemicals, warning:

- Humans and wildlife both depend on their ability to reproduce and develop normally: this is not possible without a healthy endocrine system.
- Endocrine-related diseases and disorders are on the rise worldwide.
- The speed of the increase in hormonal diseases rules out genetic factors as the only explanation.
- Thousands of laboratory studies have confirmed that exposure to EDCs contributes to endocrine disorders in humans and wildlife.
- The risk of disease due to EDCs may be significantly underestimated.
- Worldwide, there has been a failure to adequately address the underlying causes of trends in endocrine diseases and disorders.[38]

The report noted that up to 40 per cent of men in some countries had poor semen quality; the incidences of genital malformations, premature babies and neurobehavioural problems in children had all risen. 'Global rates of endocrine-related cancers (breast, endometrial, ovarian, prostate, testicular and thyroid) have been increasing over the past 40 to 50 years. The prevalence of obesity and type 2 diabetes has dramatically increased worldwide over the last 40 years,' it added.

It is not only humans who suffer from the avalanche of EDCs. Wild animals, birds, frogs, fish and even insects are having their natural systems disrupted by EDCs in much the same way. Hormone-related diseases are on the upsurge, in line with growth in output of the chemicals industry, another European report found: 'Rates of endocrine diseases and disorders, such as some reproductive and developmental harm in human populations, have changed in line with the growth of the chemical industry, leading to concerns that these factors may be linked.'[39]

An early WHO/UNEP study noted many scientific reports finding that exposure to certain EDCs has contributed to adverse effects in some wildlife species and populations. These effects varied from subtle changes in the physiology and sexual behaviour of species to permanently altered sexual differentiation. Typical examples included:

- Mammals: Exposure to PCBs and DDE damages the reproductive and immune function in Baltic seals, resulting in marked population declines.
- Birds: Eggshell thinning and altered gonads were observed in birds of prey exposed to DDT, resulting in severe population declines. Embryonic abnormalities have been observed in fish-eating birds and were directly related to PCB exposure.
- Reptiles: A pesticide spill in Lake Apopka (Florida, USA) led to a population decline in alligators linked to reproductive abnormalities.
- Amphibians: Population declines in frogs have occurred in polluted habitats worldwide.
- Fish: There is extensive evidence that chemicals from pulp mills and sewage treatment plants affect fish reproduction.
- Invertebrates: Exposure of marine shellfish to TBT (in antifouling paints) caused them to change sex, leading to population declines.[40]

Because of our common evolutionary heritage, there are strong resemblances between human hormones and those of animals, reptiles and even fish: whatever affects the delicate reproductive and growth mechanisms of wildlife may also affect us in similar ways and vice versa. The big picture now displays a Planet awash with these subtle poisons – leading to a steady increase in human infertility in all societies,[41] as well as increased rates of extinction in wildlife.

A recent ugly development in the public discussion over what to do about EDCs is the emergence of a professional counter-lobby – modelled on the same lines as climate denial – out to confuse, delay and sabotage action to protect society. In 2020, reported the French national newspaper *Le Monde:* 'A group of toxicologists with tenuous expertise and veiled conflicts of interest are working to derail the implementation of European regulations on synthetic substances that are toxic at very low doses.

They call themselves "prominent" specialists; they are not. They solemnly declare that they have no conflict of interest; however, half of them have collaborated with the chemical, pesticide, food or cosmetics industry over the last three years.' The group is 'radically opposed to any regulation of endocrine-disrupting chemicals in Europe', the paper adds.[42]

The Poisoned Generation

'Children today are sicker than they were a generation ago. From childhood cancers to autism, birth defects and asthma, a wide range of childhood diseases and disorders are on the rise. Our assessment of the latest science leaves little room for doubt: pesticides are one key driver of this sobering trend,' declared Kristin Schafer of the Pesticide Action Network (PAN).[43]

The PAN is a worldwide network founded in 1982 and connecting 600 groups in ninety countries. It explains: 'Pesticides don't respect national borders. Tons of agricultural chemicals cross international boundaries every year, either through the international marketplace or carried by wind and water currents.'[44]

Its study, *A Generation in Jeopardy*, says, 'Compelling evidence now links pesticide exposures with harms to the structure and functioning of the brain and nervous system. Neurotoxic pesticides are clearly implicated as contributors to the rising rates of attention deficit/hyperactivity disorder, autism, widespread declines in IQ and other measures of cognitive function.'

According to the report:

- Pesticide exposure now contributes to a number of increasingly common health problems for children, including cancer, birth defects and early puberty.
- Evidence connecting pesticides to certain childhood cancers is particularly strong.
- Extremely low levels of pesticide exposure can result in significant health damage, particularly during pregnancy and early childhood.

- Emerging science suggests that pesticides may also be important contributors to the current epidemics of childhood asthma, obesity and diabetes.

Not only pesticides are linked to the pandemic of childhood disorders. It is also associated with exposure to heavy metals, air pollution, poor diets, food chain contaminants, EDCs, volatile gases and vapours emitted by plastics, domestic cleaning agents, furnishings, bedding, perfumes, toys and baby products. It is the total chemical avalanche which hits them, even before they are born – an avalanche that adults can sometimes withstand, but to which infants and children are uniquely susceptible.

Babies are especially at risk of poisons because, relative to adults, they consume seven times more water and three to four times more food, and breathe twice as much air for their body-weight. They also have a lesser ability to break down and excrete chemicals. This increases their overall risk when exposed to any form of toxin. At the same time, their developing central nervous and hormonal systems are more prone to disruption by EDCs, heavy metals, volatiles and other toxins, with the possibility that early damage can last a lifetime. A trailblazing US study in 2011 by Coleen Boyle and colleagues found that one in six American children had a developmental disability and that the incidence was increasing, requiring ever-greater health and educational resources.[45] This followed hard on the heels of a series of international studies pointing to a 400 per cent rise in childhood disability over the half century from 1960 to 2010.[46] The increase was especially marked in rich societies.[47] Confirmation of the trend came in 2019 with a further study – also by Boyle and colleagues – which found that 18 per cent of US children aged 3–17 years, nearly one in every five, had a developmental disability in 2017 – an increase of almost one-third in just twenty years.[48] The leading disability by far was ADHD, followed by learning disability and autism. Indeed, the prevalence of autism more than doubled.

Rising damage to the world's children has been brought to the attention of all governments, corporations and international agencies over decades, in ways they should have been unable to ignore. Yet most of them still managed to do so.

For example, in 2006 Professors Philip Landrigan and Joel Forman told a World Health Organization conference: 'Fetuses, infants and children are exquisitely sensitive to environmental exposure (to chemicals). Recent evidence confirms children are exposed more than adults.'[49] They referred in particular to the work of Professor David Barker, of Southampton University, who argued that the early environment is a powerful determinant of a person's health for the whole of their life.

'Children are surrounded by a large and increasing number of chemicals. There is increasing evidence of toxic effects and disease,' they said. 'About 80,000 chemicals are produced commercially, and three thousand of these are produced in volumes of more than a million pounds [454,000 kilograms] a year.' No basic toxicity information is available for about half of these high-volume chemicals, the researchers added, and no information on their impact on developing children is available for 80 per cent of them. US studies confirmed that higher levels of industrial chemicals were present in children than in adults, indicating that children are more dangerously exposed.

The global mass poisoning of children by manufactured chemicals first came to world attention in the late 1950s with the singular case of the morning-sickness drug, thalidomide, which caused birth deformities in an estimated 10–20,000 children in forty-six countries. There has since been a rolling thunder of published scientific evidence of other chemical-induced effects in the very young, especially involving damage to the developing brain, nervous system, mind and reproductive system.

Landrigan and Forman added:

Initial recognition was of obvious, massive damage caused by high-dose exposures . . . But later studies with more sophisticated tools have shown in every case that the brain injury caused by toxic chemicals is not limited to obvious conditions. It is now recognized that there exists an entire spectrum of diminished brain function in persons exposed to toxic chemicals, termed subclinical toxicity.

Widespread subclinical neurotoxicity can affect the health, well-being, intelligence and even the security of entire societies.

They pointed out that the loss of five IQ points per child would increase society's number of mentally retarded people by around two-thirds. They warned that the impact could be lifelong, adding: 'Early exposure to toxic chemicals may increase risk of degenerative brain disease in later life, such as Parkinson's disease and Alzheimer's disease.' The researchers cautioned that the lion's share of the current ongoing economic cost of childhood diseases, which runs into many tens of billions of dollars, is due to reduced intelligence and lowered productivity: 'Environmentally attributable disease is *very costly to society*' (source italics).

The following sections deal with specific aspects of the universal poisoning of the world's children.

Not Paying Attention

Attention deficit hyperactivity disorder (ADHD) is today among the most common psychiatric or behavioural problems, affecting around 375 million children and adults globally.

While some forms of childhood disability have been declining, those affecting the brain, central nervous system, intelligence and behaviour have risen significantly in recent decades. One estimate puts the current prevalence of these conditions at about 7.2 per cent of the world's children – equal to about 175 million individuals.[50] Other estimates put the ADHD incidence at 8.4 per cent for children and 2.5–4 per cent for adults.[51] The US Centers for Disease Control estimated that American ADHD prevalence rose from 7.8 per cent in 2003 to 11 per cent in 2012 (an increase of 42 per cent over nine years) among children and is around 4 per cent in adults. Furthermore, it found the condition was twice as common among boys (15 per cent) as among girls (7 per cent).[52] About half of children who suffer from ADHD continue to show signs of it in adulthood. Around two-thirds of children diagnosed with ADHD also show other mental conditions, including autism, depression, emotional extremes, anxiety, twitching (Tourette's) and drug abuse.

The global picture is complicated by varying diagnostic and medical treatment criteria in different countries.

ADHD is thought to be partly an inherited condition (about half of families with an affected child have one parent with the condition), while its diagnosis has certainly improved during recent decades. However, these factors alone do not account for the rapid increase in its global prevalence, in developed industrial societies or in adults. As with tobacco and cancer, direct links are hard to prove – but it is also difficult to overlook the fact that a serious brain condition has burgeoned over the same timeframe as an exponential increase in neurotoxins and hormone disruptors in the child's home and food.

There are three forms of ADHD: inattentive, hyperactive/impulsive and the two combined. In the first case the main symptoms are inattention, easy distraction, not listening when addressed, being disorganised, constantly losing things and being forgetful of daily tasks. In the second, the signs are inability to sit still or be quiet, constant fidgeting or standing, running about, talking too much, cutting other people off, impatience and intrusive behaviour. There is no laboratory test for ADHD: diagnosis is based purely on observed behaviour. There is no cure, although the condition can be reduced by treatment and may abate naturally.

Moreover, ADHD is not a condition that harms only those affected: the World Health Organization lists adult ADHD – now affecting as many as 200 million people – as a key risk factor for acts of murder and violence, drug use, traffic violations and suicide.

The most alarming attention deficit on the Planet is the lack of attention to the issue paid by governments, corporations and their shareholders – who are the economic beneficiaries of this pandemic of child poisoning. As a result, ADHD is treated as individual cases, rather than the universal condition it has become. The whole focus is on enabling children to develop ADHD and then seeing whether they can be treated or not, rather than attempting to prevent the condition in the first place. Where prevention is recommended, the responsibility is almost invariably, and very unfairly, placed on the mother. Governments, as a general rule, accept little responsibility for preventing mind damage to young generations, nor do they enforce controls upon industries that produce neurotoxins, nor penalties on the investors who make money from this form of child abuse.

***

As every worried parent knows, autism is on the rise – not just in the West, but all around the world, and especially in newly industrialising countries such as China and India. Autism describes a range of mental conditions (including ADHD) that affect the individual's behaviour and their ability to communicate and interact with others – the term means 'self-focus'. Isolated cases are recorded from the eighteenth century, and the condition was first described in the medical literature in 1943.

The World Health Organization estimates autism's global prevalence at one in every 160 children – about 16 million individuals.[53] In the USA, cases of autism spectrum disorder (ASD) nearly tripled between 2000 and 2020 and were three times the world average.[54] To illustrate the trend:

- in the 1950s it is estimated that 1 in 25,000 US children was diagnosed with autism;
- in the 1970s and 1980s, about 1 in 2500 children was diagnosed with ASD;
- in 2000, the US Centers for Disease Control (CDC) reported that ASD affected 1 in 150 American children;
- in 2004 this figure had reached 1 in 125 children;
- in 2006 the figure was 1 in 110 children;
- in 2008 this figure reached 1 in 88, based on the CDC's ADDM (Autism and Developmental Disabilities Monitoring) network of fourteen monitoring areas across the USA;
- in March 2013, the US National Health Statistics Report indicated that 1 in 50 children across the USA had been diagnosed with ASD;

- in heavily populated US cities this subsequently reached 1 in 27 children; surveys of parents indicated the rate may be as high as one child in nine.

A thousandfold increase in autism prevalence, from the 1950s to the 2020s, is hard to explain in terms of improved diagnosis alone, and even less so in terms of genetics. Something profound has clearly changed in the child's growing environment and, since the symptoms of autism can be diagnosed as early as the first 12–18 months of a baby's life, it is probably not something like overuse of smartphones or poor education, as some experts have speculated.

The triggers of this sudden and dramatic increase in autism have not yet been pinpointed but, like the suspicions surrounding lead in the 1970s, a rising mountain of science points with growing certainty at various forms of industrial, domestic and food pollution that can damage the child's developing brain. As Harvard professors Grandjean and Landrigan cautioned: 'Two hundred ... chemicals are known to cause clinical neurotoxicity in adults. Despite a lack of systematic testing, many more are known to be neurotoxic in laboratory models. Their toxicity to the developing human brain is not known and they are not regulated to protect children.'[55] In a subsequent study in 2015 they identified ten specific brain poisons: lead, methylmercury, polychlorinated biphenyls, arsenic, toluene, manganese, fluoride, chlorpyrifos, dichlorodiphenyltrichloroethane (DDT), tetrachloroethylene (TCE) and the polybrominated diphenyl ethers (PBDE).[56] The researchers called for a worldwide effort to prevent brain damage of children: 'we propose a global prevention strategy. Untested chemicals should not be presumed to be safe to brain development, and chemicals in existing use and all new chemicals must therefore be tested for developmental neurotoxicity.'

Adding weight to the idea that babies are immersed in neurotoxins from birth, researchers at the University of Southern California found evidence of a link between child autism rates and mothers who lived close to a freeway during late pregnancy.[57] Since diesel fumes alone contain at least 650 different chemicals and particulates and they in turn are but part of total highway emissions, the task of unravelling which particular substances may or may not be responsible for conditions such as autism is extraordinarily difficult and may take years, as it did with tobacco smoke.

In 2012 an investigation by Harvard School of Public Health involving more than 110,000 nurses across the USA concluded: 'Perinatal exposure to higher levels of diesel, lead, manganese, mercury, methylene chloride, and a combined measure of metals were linearly related to increased risk of autism.' Essentially it, too, found that women living in the most highly polluted areas of the country were twice as likely to have an autistic child.[58] It was another smoking gun. Several subsequent studies have reached similar conclusions.

Scientists have long known that urban air pollution is implicated in much besides autism – cancers, heart disease, lung disease and asthma, cystic fibrosis, genetic and other developmental disorders – and that it kills around 8 million people worldwide every year. Clean air laws have brought some improvement in certain cities, but over much of the Planet the combination of industrial smog, especially in Asia, and smoke from wildfires fanned by climate change are making world air quality markedly worse. It is a form of poisoning to which nine people in every ten are now exposed, says the World Health Organization.[59]

***

Rising rates of childhood cancer are one of the most distressing manifestations of our age: while actual numbers remain relatively small, cancer is nevertheless a leading cause of death for children and adolescents around the world, and approximately 300,000 children aged 0–19 years old are diagnosed each year. The most common categories include leukaemias, brain cancers, lymphomas and solid tumours, such as neuroblastoma and Wilms' tumour. In rich countries four out of five children with cancer recover, but in many low- and middle-income countries only about one in five survives.[60]

It is frequently stated that 'The causes of most childhood cancers are unknown' – since the trigger is generally only traced in 5 to 10 per cent of cases. However Belgian researchers Nicholas van Larebeke and colleagues argued that industrial chemicals must be seriously considered among the possible risk factors:

> Children are at risk of exposure to over 15,000 high-production-volume chemicals and are certainly exposed to many carcinogens. The individual impacts of most of these agents are too small to be detected, but collectively these unrecognized factors are potentially important. Infants and children are exposed to higher levels of some environmental toxicants and may also be more sensitive. During intrauterine development and childhood, cells divide frequently, and the mutant frequency rises rapidly. Endocrine-related cancers or susceptibility to cancer may result from developmental exposures rather than from exposures existing at or near the time of diagnosis.[61]

As in the case of brain diseases, air pollution is a primary suspect. A Canadian study found ambient air pollution in pregnancy increases the risk of childhood cancers, that exposure to microparticles in polluted air in the first trimester increased the risk of astrocytoma, and first trimester exposure to nitrogen dioxide increased risk of acute lymphoblastic leukaemia.[62]

Childhood cancer rates are generally not increasing very fast – around 0.6 to 1.1 per cent a year in North America and Europe, and somewhat faster in Asia and Africa. Typifying the trend in developed countries, an Australian study found the rate of childhood cancer there had increased by 1.2 per cent each year between 2005 and 2015, and was expected to rise at least a further 7 per cent by 2035.[63]

However, the fact that childhood cancers are increasing at all suggests that the triggers – whatever they may be – are also becoming more widespread. No efforts should be spared to pin these down, however difficult or costly it may prove, and eliminate them. In the case of childhood cancer especially, prevention is infinitely more ethical, humane and cost-effective than cure. However, far fewer scientific and medical resources are devoted to prevention rather than to discovering 'cures'. The latter are still only partly effective, and one has to wonder why the focus of medical research funding has been on treatment rather than on prevention – and about the degree of influence exerted on the research system by large pharmaceutical corporates seeking profitable new drugs.

Asthma Attack

Asthma is another disease on the rise in the human population that has been linked to various chemical triggers. The WHO estimates about a third of a billion people suffer from asthma, of whom around 420,000 die from it each year.[64] While the primary cause is probably genetic, the condition is also brought on by exposure to irritants such as household dust, allergens, tobacco smoke, air pollution and certain airborne chemical vapours. The disease is increasing among children in particular.

The increase is attributed in a wide range of scientific studies to the rise in exposure to chemical lung irritants and polluted air both in the home and in large cities, and to soil erosion and airborne dust particles in rural areas.[65] For example, a Chinese study found '… a "modern" home environment together with a modern lifestyle is associated with increased prevalences of asthma and allergies among children. Appropriate indoor environmental interventions and education of children's caregivers are important in the management of childhood asthma and allergy.'[66]

The US Centers for Disease Control says more than 25 million Americans have asthma. This includes 7.7 per cent of adults and 8.4 per cent of children. Asthma rates have been increasing since the early 1980s in all age, sex and racial groups and it is now the leading chronic disease of US children.[67] Death rates are ten in every million, and have been trending down due to better emergency care.[68]

Mystery Plagues

The number of previously unknown and unexplained human conditions is multiplying. Some studies indicate that up to 45 per cent of all doctor's appointments and half of hospital visits deal with medically inexplicable symptoms.[69]

Recent examples include multiple chemical sensitivity (MCS), Gulf War syndrome, stiff person syndrome, Morgellon's disease, cycle vomiting syndrome (CVS), electromagnetic hypersensitivity, chronic fatigue syndrome (CFS), irritable bowel syndrome (IBS), fibro- and polymyalgia and attention deficit hyperactivity disorder (ADHD).

Many of these conditions remain controversial and are generally classified by doctors as MUCs, or 'medically unexplained conditions' – or else are not recognised as diseases at all. For patients with these syndromes, one of the most distressing features of their experience is that doctors frequently attribute the symptoms to a psychiatric disorder, without exploring what in the patient's life and living environment might be leading to their symptoms and whether they are real or imagined. 'Clinicians have developed a range of strategies for deflecting the threat to medical competence posed by medically unexplained symptoms. Generally, these involve shifting the blame from the limits of medicine to some characteristic of the patient. Given our psychologically oriented culture, it is an easy slide from declaring a symptom unexplained to attributing it to specific psychological traits or states of the patient,' Kirkmayer and colleagues explain in a paper on MUCs.[70]

This 'blame the patient' mentality on the part of some doctors may, in part, explain why society has been slow to explore the evidence for chronic chemical poisoning. The tendency has been to encourage patients to blame themselves – either their genes, their imaginations or their mental stability – rather than to mount rational, evidence-based inquiry into other possibilities, such as the patient's living environment and exposure to the multiplicity of toxins present in it. This is not to say that all of these mysterious conditions are linked to chemicals, but rather that the possibility is firming that some MUCs may in fact be linked to our immersive daily exposure to thousands of tiny doses of toxic man-made substances, to which our ancestors were never exposed. Ethics and logic dictate that we should assess this possibility with the same degree of care, good science and open-minded investigation as for any other possible source of injury.

Morgellon's disease offers an intriguing example. According to the Mayo Clinic, 'Morgellon's disease is the popular name for an unexplained skin disorder characterized by disfiguring sores and crawling sensations on and under the skin. Morgellon's disease also features fibers or solid materials emerging from these sores.' The explanation continues: 'Fibers found in the sores are usually wisps of cotton thread, probably coming from clothing or bandages. CDC experts note that the signs and symptoms of Morgellon's disease are very similar to those of a mental illness involving false beliefs about infestation by parasites (delusional parasitosis).' Their only clinical advice is 'Get treatment for anxiety, depression or any other condition that affects your thinking.'[71] So, in the Mayo's view, Morgellon's is probably due to a mental state – but if so, what was the trigger for that? And can we rule out an external neurotoxin affecting either skin, brain or both?

Tens of thousands of Vietnam, Gulf and Afghanistan war veterans would recognise such dismissive responses from the medical establishment. Nobody wanted to believe the veterans either, when they argued that their cancers, depression and deformed children might be linked to chemical exposure during their military service. Decades on, and with the powerful evidence of doctors working with Vietnamese victims, which strongly supports the theory that herbicides and other toxic chemicals can cause health havoc in people, the prevailing view has changed: while officialdom may continue to quibble, both laws and lawsuits are increasingly settled in the veterans' favour, indicating a general acknowledgement it wasn't their imaginations. For example, exposure to herbicides used in the Vietnam War has now been medically associated, among others, with: AL amyloidosis, leukaemia, Hodgkin's and non-Hodgkin's diseases, heart disease, chloracne, type 2 diabetes, Parkinson's disease, respiratory cancers, prostate cancer, soft tissue sarcoma and multiple myeloma.[72]

For civilians worldwide exposed to similar chemicals and suffering similar conditions, recognition remains largely in abeyance.

Old Plagues Ascendant

Over recent years there has been a marked increase in conditions long-recognised by medicine, but previously seen at much lower levels in the population – depression, schizophrenia, learning difficulties, Parkinson's disease, Alzheimer's disease, other mental disorders, autoimmune diseases, asthma, obesity and diabetes – most of which are now medically tied with exposure to man-made chemicals. For example, researchers Anthony Wang and colleagues found California farm workers with high exposure to various common pesticides were three times more likely to suffer Parkinson's disease, particularly the early onset variety.[73] Similarly, a team led by Kathleen M. Hayden of Duke University studied 3000 elderly people in Cache Country, Utah, and found a correlation between pesticide exposure and higher rates of both dementia and Alzheimer's. 'Pesticide exposure may increase the risk of dementia and Alzheimer's disease in late life,' they concluded.[74] The WHO says there are currently around 50 million people living with dementia, and 10 million new cases are being diagnosed each year. At such rates, the world will have a 'nation of the demented' 152 million strong by the mid-century.[75]

Public awareness of the perils of pesticides has led to an upsurge in legal actions against offending companies. A major case was the lawsuit surrounding the world's most popular herbicide, glyphosate. By early 2020 more than 125,000 users of the herbicide, mostly farmers, gardeners and their families, had sued the owner of glyphosate, German chemical giant Bayer, which had acquired it when it bought out the developer, Monsanto, in 2018. The plaintiffs cited ten different cancers – nine lymphomas and a leukaemia – which they attributed to glyphosate exposure. In June 2020, after its shares had taken a pummelling, Bayer offered to settle the bulk of cases for a combined $11 billion.[76] The pebble that started the legal avalanche was rolled by a Californian school groundskeeper, Lee Johnson, who in 2014 at the age of forty-six was diagnosed with terminal cancer after using the herbicide an average of thirty times a year for much of his working life. The trial was accelerated because Johnson did not have long to live.

In an interview with *Time* magazine, Johnson recounted:

> *Before I got sick life was pretty good. My job title at the school district was integrated pest manager, IPM. I did everything – caught skunks, mice, and raccoons, patched holes in walls, worked on irrigation issues. And I sprayed the pesticides. I had to be at work by sun-up to make sure we had time to spray before the kids got to school. On a typical day I would fill up my little container with raw pesticide liquid and then put that in the back of my truck and then mix a load before I would leave the yard. I did not like using the chemicals but I loved that job.*

> *That day of the accident, the day the sprayer broke and I got drenched in the juice . . . I washed up in the sink as best I could and changed my clothes. Later I went home and took a good long shower. Then I got a little rash. Then it got worse and worse and worse. At one point I had lesions on my face, on my lips, all over my arms and legs.*

Doctors puzzled over his symptoms and ran tests until, one day, they called him in to tell him he had cancer. Johnson was a modest person who never mentioned the herbicide company by name and insisted he didn't even want an apology. He just wanted to know why his health had suddenly failed. And he wanted to help other schools avoid a repeat.[77]

Among the evidence laid before jurors at the trial were internal Monsanto documents that referred to plans to 'ghostwrite' scientific papers claiming its products were safe and which admitted the company had no proof its products did not cause cancer.[78] Also revealed were scurrilous attempts to discredit the WHO's International Agency for Research on Cancer (IARC), which had determined glyphosate to be a 'probable human carcinogen'.[79] Persuaded by the evidence, the Californian jury awarded Johnson an astonishing $289 million, which the Judge reduced to a more modest $78 million.

Behind the scenes a power struggle was being played out. The US Environmental Protection Agency, having originally concluded from animal trials that America's favourite herbicide was probably a carcinogen in 1986, reversed itself in 2020, declaring 'there are no risks of concern to human health when glyphosate is used according to the label and that it is not a carcinogen'.[80] In Europe public opposition to glyphosate had been building for years after the chemical was increasingly found in food. In 2016 a poll showed two-thirds of respondents favoured a European ban on glyphosate. In 2017, over 1.3 million people signed a petition demanding a ban. In 2019 a European Parliament report found that EU regulators had plagiarised their decision to relicense glyphosate from documents prepared by pesticide companies, including Monsanto. The scandal led to several countries immediately banning or restricting the chemical. The Danish Working Environment Authority declared glyphosate a carcinogen.[81]

The outcome echoed the earlier divide between Europe and the USA over other farm chemicals and food additives, with American regulators siding with corporations over citizens, while Europe took a more precautionary course. Meanwhile Bayer, despite its huge legal payout, declared it would continue to sell the herbicide and would not be adding a cancer warning label to the product.

Another condition known to medicine for many years, formerly rare but now on the rise globally, is autoimmunity which currently affects around 5 per cent of the population – 400 million people – according to the World Health Organization. Autoimmune disorders include type 2 diabetes, lupus, inflammatory bowel disease, rheumatoid arthritis, multiple sclerosis, psoriasis, Guillain–Barré syndrome, Graves' disease and Crohn's disease. These top the list of eighty known conditions that involve the body mounting an immune-based attack on its own tissues. The British Society for Immunology warned the UK Parliament in 2018 that 4 million Britons – one in sixteen – were now living with an autoimmune disorder, often with more than one. It warned that these diseases were 'under-reported' and urged a combined research effort to tackle the problem of causes and prevention.[82] The causes of autoimmune disease are still not clearly understood: firstly, the patient probably has a genetic susceptibility which makes them vulnerable, secondly, something in their living environment scrambles the body's signalling system, causing it to attack itself. The key question is: what triggers such attacks? Theories include microbes such as streptococcus, certain medical drugs, tobacco smoking, certain foods such as wheat, and certain chemicals including silica, aniline dyes and solvents such as TCE. The evidence that there is a trigger in the external environment of the victim is strong, according to the US National Institutes of Health[83] – but sorting out just which substance or microbe is the trigger for which disease is the hard part. Genetics alone cannot explain why autoimmune disorders are ratcheting up globally. However, as with other diseases, the rise in their prevalence parallels the dramatic increase in chemical use in the human environment over the last fifty years. Autoimmunity may well be another smoking gun.

Cancerous Growth

Cancer has become the Killer of our Age, in both the developing and developed worlds. The condition costs around $1.5 trillion a year to diagnose, treat and in terms of lives and work lost. The World Health Organization reported 18 million new cases in 2018, along with 10 million fatalities or one in six of all deaths.[84] The number of new cases a year is forecast to reach 30 million by 2040.[85]

Overall, the rate at which people are developing cancers appears to be increasing at around 2.5 per cent per year. In developed countries, one in every three women and one in every two men will be diagnosed with a cancer at some point in their lives. World cancer rates are 200 in every 100,000 people, but in the worst-affected countries – Australia, New Zealand, Ireland, Hungary and the USA – may be twice that.[86] A significant part of the rise in cancer is due to the ageing population – older people being more prone to cancers – and also to success in treating infections and other lethal conditions, leaving cancer as the

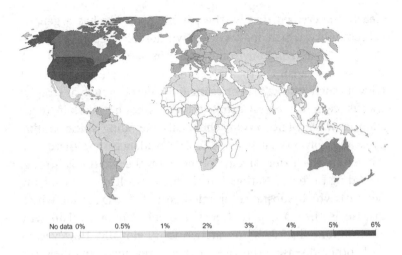

No data 0%   0.5%   1%   2%   3%   4%   5%   6%

Figure 6.2. World cancer rates, by country, 2017. Source: Our World in Data, IHME.

remaining great reaper. In developed countries success in treatment has doubled survival chances, with about one in two patients living for ten years or more following treatment. However, it is widely acknowledged that fixing cancer now depends on widespread prevention, as much or even more than on merely extending the lives of cancer victims. Cancer is primarily a disease of affluence, as Figure 6.2 indicates.

'Despite exciting advances ... we cannot treat our way out of the cancer problem,' Dr Christopher Wild, Director of the International Agency for Research on Cancer (IARC), commented in the 2014 World Cancer Report. 'More commitment to prevention and early detection is desperately needed in order to complement improved treatments and address the alarming rise in cancer burden globally.'[87]

While tobacco smoking, alcohol, obesity and poor diet (all of which have chemical exposure components) are among the main risk factors for contracting a cancer, chemical exposure in the home, workplace or urban environment is also now recognised as a major risk. The IARC has, over more than four decades, amassed databases on known and suspected carcinogens (cancer-causing substances); as of 2020 these contained

120 known carcinogens, 88 probable carcinogens, 315 possible carcinogens and 499 substances that had not yet been classified.[88] These are associated with cancers of practically every part of the human body.[89] Of the remainder, 349,000 chemicals have yet to be tested.

The US Agency for Toxic Substances and Disease Registry (ATSDR) warns: 'Exposure to chemicals in the outdoors, at home, and at work may add to your chances of getting cancer.' But offers little in the way of practical advice on how to avoid it, other than to quit smoking. It lists known carcinogens such as asbestos, arsenic, benzene, beryllium and vinyl chloride. Possible carcinogens are substances that cause cancer in animals and probably also do so in humans: they include DDT, chloroform, formaldehyde and polychlorinated biphenyls (PCBs). 'A person's risk of developing cancer depends on how much, how long, how often, and when they are exposed to these chemicals. When you are exposed is important because a small exposure in the womb, for example, may be more serious than a small exposure as an adult,' the Agency cautions.[90] Science recognises three main types of carcinogens:

- chemicals that can cause cancer (direct carcinogens),
- chemicals that do not cause cancer unless they are changed when they are metabolised in your body (procarcinogens), and
- chemicals that do not cause cancer by themselves but act with another chemical to cause cancer (cocarcinogens).

Only about 5–10 per cent of cancers are thought to be entirely heritable,[91] the rest having some external trigger – although individuals may have an inherited susceptibility. The World Health Organization states that from a third to a half of all cancers are preventable, by reducing our exposure to the main known causes.[92] Many of those known causes involve occupational or industrial exposure (Table 6.1).

Breast cancer is the commonest cancer in women, with 2 million new cases and over half a million deaths each year. Its incidence is rising partly because women are living longer lives and carrying more weight – and partly due to other factors,

Table 6.1 Examples of cancers linked to occupational exposure to
various chemicals

Cancer	Substance, source or activity
Lung	Arsenic, asbestos, cadmium, coke oven fumes, chromium compounds, coal gasification, nickel refining, foundry substances, radon, soot, tars, oils, silica, traffic fumes
Bladder	Aluminium production, rubber industry, leather industry, 4-aminobiphenyl, benzidine
Nasal cavity and sinuses	Formaldehyde, isopropyl alcohol manufacture, mustard gas, nickel refining, leather dust, wood dust
Larynx	Asbestos, isopropyl alcohol, mustard gas
Pharynx	Formaldehyde, mustard gas
Mesothelioma (lung)	Asbestos
Lymphatic and hematopoietic system	Benzene, ethylene oxide, herbicides, X-radiation
Skin	Arsenic, coal tars, mineral oils, sunlight
Soft-tissue sarcoma	Chlorophenols, chlorophenoxyl herbicides
Liver	Arsenic, vinyl chloride
Lip	Sunlight

Sources: American Cancer Society; IARC.

including exposure to 'a cocktail of carcinogens and endocrine
disruptors every day that puts us at greater risk' in food, cosmetics, personal care products, furnishings and household cleaning
products, as Breast Cancer Fund CEO Jeanne Rizzo put it.[93]

Rizzo was a member of the US Congress-appointed
Interagency Breast Cancer and Environmental Research
Coordinating Committee, which in 2013 called for a national
breast cancer prevention strategy, along with a major increase in
research efforts to pinpoint the causes. The report noted that
fewer than 2 per cent of chemicals used in the USA had been
tested for their ability to cause cancer.[94] The breast cancer
picture is complicated by the fact that endocrine-disrupting

chemicals (EDCs) also appear implicated in the process of starting a tumour, potentially adding a further 1400 substances to the list of known carcinogens. Furthermore, numerous studies have found links between breast cancer risk and things such as diesel exhaust fumes, night shift-work, radiation and occupational exposure in industries such as farming, horticulture, shoemaking, canning, metal coating, telecommunications, dry cleaning and pest control.[95]

Survival rates from cancer of the breast, bowel and other organs are improving in rich countries. This affluence is ironically thanks to drugs made by the petrochemical and pharmaceutical industries – industries whose other products are increasingly suspected of causing a large part of the problem in the first place. This raises the disturbing scenario that, the more effective the number of cancer treatments, the less urgent it becomes to find and prevent chemicals that cause cancers in the first place. Chemotherapy may thus prove a seductive highway to higher rates of the disease.

Today, thanks to treatments, **cancer is at risk of becoming the acceptable price of living in a chemicalised society.**

A Melancholy Tale

Depression, or melancholia as it was once known, is a condition of the mind that has been recognised for over two thousand years, since Classical times, but whose prevalence has increased sharply in the modern world. The World Health Organization estimates there are more than 350 million sufferers globally – 4 per cent of the population; the majority of cases are untreated and many end in suicide, making it the tenth most common cause of death and second most common cause of disability (after heart disease).[96] Some experts consider as much as 10 or even 20 per cent of the population may be affected. Between 2005 and 2015, the number of people suffering depressive illness rose by nearly 20 per cent. People born after 1945 are ten times more likely to suffer depression than their forebears.[97] Depression rates are similar across countries and cultures, but the USA is the world's most depressed society, with one person in twenty diagnosed.

As we have seen with the issue of intelligence, the human brain is an organ that depends on delicate chemical signalling to function and is thus exquisitely sensitive when exposed to chemicals, natural and man-made, at minuscule concentrations. Our enthusiastic use of alcohol, tobacco and recreational drugs derives from our desire, sometimes our compulsion, to artificially tweak the pleasure centres of our brain chemically.

Depression has long been described as 'a chemical imbalance in the brain', but this statement really explains nothing, including what causes it, and has been dubbed by some a myth.[98] Harvard Medical School explains:

> Research suggests that depression doesn't spring from simply having too much or too little of certain brain chemicals. Rather, there are many possible causes of depression, including faulty mood regulation by the brain, genetic vulnerability, stressful life events, medications, and medical problems. It's believed that several of these forces interact to bring on depression. To be sure, chemicals are involved in this process, but it is not a simple matter of one chemical being too low and another too high. Rather, many chemicals are involved, working both inside and outside nerve cells. There are millions, even billions, of chemical reactions that make up the dynamic system that is responsible for your mood, perceptions, and how you experience life ...[99]

Science has established that most depressive illness is not genetic but acquired – and that it involves some malfunction of the brain's signalling system involving neurotransmitters such as serotonin, norepinephrine and, possibly, dopamine.

Evidence that man-made chemicals are implicated in the chain of events that cause the natural signalling in our brain to malfunction, resulting in depression, is accumulating – but the exact process, and which chemical triggers, are far from clear. Nevertheless, the ugly fact that the world's citizens are today ten times more depressed than the World War 2 generation while the volume and range of chemical exposures has exploded – many of them to known neurotoxins – suggests more than mere coincidence. A major study in the *Annual Review of Public Health* stated, 'chemical pollutants, may also have a neurobiological impact and influence the risk of depression, especially among genetically susceptible individuals'.[100]

Apart from stress or sorrow in family, work or personal life, depression has also been linked with external causes such as poor diet, lack of vitamin D or folate, certain medications, alcohol and drug abuse. One suggestive case involved a long-term study of farmers and their wives on farms using organophosphate pesticides, which are known to cause neurological damage: US researchers concluded, 'exposure to pesticides at a high enough concentration to cause self-reported poisoning symptoms was associated with high depressive symptoms independently of other known risk factors for depression among farm residents'.[101] In another case, more than half the residents of 1800 homes in a small Mississippi town who were exposed to pest control chemicals over ten years were diagnosed as depressed years after the event.[102] A third study found strong links between pesticide exposure and depression among elderly Korean farmers – with those who were more exposed suffering worse depression.[103] A review of 167 reports into farmers' mental health worldwide concluded it was, typically, poorer than for city people and that the four most common factors were pesticide use, financial difficulties, climate/weather impacts and poor physical health.[104] Together, these put pesticides squarely in the gunsight as a potential major cause of depression.

The fact that certain medical drugs – including some anti-depressants![105] – also produce depression as a side-effect lends weight to the view that depression can be a chemically induced illness, and that our greatly increased exposure in the modern age is implicated in its upsurge.

There remains, as yet, insufficient research to answer the question 'How much of the depression suffered by today's society is attributable to man-made chemicals?' Answering it is urgent: beside the personal suffering and economic loss it

causes, depression is implicated in many of the 1 million suicides and 20 million attempted suicides that occur each year,
according to the WHO. Like cancer, depression may well be
another largely preventable pandemic.

Chemical Obesity

The world is in the grip of a girth pandemic, many times worse
and far more deadly than Covid-19 or any of the infectious
diseases the public generally recognises as universal plagues.

The World Health Organization points out that the global rate
of obesity has tripled since 1975. More than 1.9 billion adults are
overweight and, of these, 650 million are obese. About 380 million children are overweight or obese. The condition now
affects about 30 per cent of the human population, making it
one of the worst pandemics in all history,[106] comparable to the
Black Death.

Obesity is the world's fifth leading risk factor for early death
and is rapidly overhauling starvation as a killer. It is no longer
mainly concentrated in the affluent countries but has spread
throughout the world. Diabetes, to which obesity is strongly
linked, is also increasing rapidly, with 422 million cases causing
around 1.6 million deaths a year.[107]

'It is a commonly held view that obesity is all to do with too
many calories taken in and too few expended in exercise, with a
genetic predisposition in some individuals. However, a new line
of research suggests that exposure to certain man-made chemicals in our environment can play an important role in the
development of obesity. While obesity is a known risk factor
for diabetes, evidence is growing that chemical exposures are
also implicated in diabetes,' two eminent medical scientists,
Spaniard Miquel Porta and South Korean Professor Duk-Hee
Lee, warned in a review article.[108]

Unborn babies, infants and children, especially, are at risk of
having their normal growth patterns distorted by early exposure
to EDCs, contributing to a nearly fourfold increase in global
childhood obesity, argues a more recent study.[109] This one

implicated bisphenol A (BPA), phthalates, triclosan and perfluoroalkyl substance (PFAS) flame retardants as possible sources of obesity.

Obesity and diabetes both exhibit the 'blame the patient' mentality that still lingers in parts of the medical establishment, which haughtily implies the sufferer is guilty of overeating and laziness, rather than bothering to inquire into more complex and subtle explanations for the drivers of these modern pandemics – and so risk offending powerful commercial interests. While the lifestyle and dietary choices of individuals – influenced by the torrent of unhealthy food advertising to which they are daily subjected – no doubt play a role in the ill-health and death of millions, many scientists are now convinced that other factors are also at work.

Porta and Lee, for example, say that there is plenty of evidence that chemicals found in food and the human environment cause weight gain in laboratory animals. Suspect substances include POPs, PCBs, pesticides, flame retardants, BPA, phthalates, lead, nicotine, diesel exhaust and medical drugs – all of which are also linked to a host of other conditions. 'It is likely that there are other chemicals in the environment that increase the risk of obesity, which have yet to be recognised,' they add. The biochemical mechanisms involved are not yet clear, but probably include confounding the body's natural energy storage and distribution system and damaging its genes. The latter raises the disturbing implication that conditions such as obesity and diabetes caused by exposure to synthetic chemicals may now be inherited by our children through the genetic damage they cause. They conclude, '. . . the concern that chemicals in the environment may be partly responsible for the increasing occurrence of obesity in human populations is based on a significant and growing number of mechanistic studies and animal experiments, as well as on some clinical and epidemiological studies. The weight of evidence is compelling.' The growing conviction that the fatness pandemic is partly down to these substances has even given rise to a name for them: obesogens.

If chemicals are partly to blame for obesity, then they are also likely to be partly responsible for the universal upsurge in diabetes, Porta and Lee add. 'Evidence suggesting a relationship between human contamination with environmental chemicals and the risk of type 2 diabetes has existed for over 15 years, with the volume and strength of the evidence becoming particularly persuasive since 2006 ... Chemicals linked to type 2 diabetes in human studies are POPs (including dioxins, PCBs, and some organochlorine pesticides and brominated flame retardants), arsenic, BPA, organophosphate and carbamate pesticides, and certain phthalates.' Their overriding conclusion was that, 'given the current epidemics of obesity and diabetes, action to reduce exposures to many chemicals possibly implicated in obesity and, more certainly, in diabetes, is warranted on a precautionary basis.'

Since the idea of obesogens (chemicals that make you fat) was first put forward in the early twenty-first century, accumulating scientific evidence has only strengthened that opinion. Blumberg and Egusquiza, for instance, stated in 2020, 'The incidence of obesity has reached an all-time high, and this increase is observed worldwide. There is a growing need to understand all the factors that contribute to obesity to effectively treat and prevent it and associated comorbidities.'[110]

They explain that obesogens have many effects on the body's natural self-regulation: they alter our hormonal metabolism so we deposit more fat, they change our satiation response to food, they alter our gut bacteria to ones that favour our being fatter, they confuse the genes that control the number and type of fat cells deposited and, alarmingly, they change how the genes inherited by our children behave, making them fatter. 'Although the environmental obesogen field is just 15 years old, it is becoming clear that chemical exposures may be important contributors to the obesity pandemic. However, we have only just scratched the surface and need to learn much more about the number of obesogens that exist, how they act, and how we can best protect ourselves and future generations from their harmful impacts,' they said. The development of screening tests to find out which EDCs occur most often in overweight individuals is such a step.

The core problem with obesogens, like the problem with man-made chemicals in general, is that they never arrive one at a time but, rather, in complex mixtures of hundreds of substances that may have cascading effects on the victim, short and long term. Furthermore, if the testing of modern chemicals for their ability to cause cancer is limited to 2 per cent or less of the chemicals we use, then testing of chemicals for their ability to kill through obesity and diabetes is practically zero, and no government yet requires it of industry.

Consequently, it will be decades before governments are able to regulate the full suite of fat promoters. It will be left to the individual to try, as best they can, to control their own chemical intake – in the absence of objective advice and information about what's in their food and what it does to them.

Being fat may look like a simple issue to some, but it is another example of the immense complexity that the uncontrolled human chemical avalanche has unleashed.

Lethal Legacy

Society has known for millennia that too much of a toxic substance makes you sick and may even kill you. What is comparatively new is the idea that it may kill your grandkids too.

The burden of chemically induced ill-health is now in the process of transfer from one generation to the next by epigenetic means. For years, scientists considered that our genes were stable entities, unchanging from generation to generation. Since 1975 they have encountered increasing evidence that when genes are chemically damaged or silenced (disabled), their behaviour (or 'expression') changes and this can cause disease, not just in those who own the genes, but in those who inherit them too. This paradigm shift in our understanding of genetics and heritability has so far taken over four decades to sink in.

Put simply, the current view is that the basic structure of the affected gene remains apparently the same – but the biochemical commands it issues to the rest of the body can be distorted, causing disease. And your children and grandchildren risk inheriting the damaged gene and suffering from the same ailments.

Substances now implicated in causing epigenetic damage include cigarette smoke, alcohol, phthalates, BPA, metals (such as chromium-6, arsenic and methylmercury), air pollution, endocrine disruptors and dioxins. Furthermore, this damage appears to pass from parent to child. 'There are now many examples of epigenetic inheritance through the germ line,' said Australian scientist Robin Holliday, an expert on the ageing process, who pursued the heritability of epigenetic defects since the mid-1980s.[111]

Among the more striking examples of epigenesis is that a mother's diet during pregnancy can affect the risk of her child becoming obese, and that as more and more parents become overweight, the risk of inherited obesity compounds with each succeeding generation.[112] In general, this suggests that chemical damage to her own genes sustained by the mother may potentially lead to her unknowingly reprogramming the DNA of her unborn child.

An international team of scientists summed up the situation thus:

> More than 13 million deaths every year are due to environmental pollutants, and as much as 24% of diseases are estimated to be caused by environmental exposures that can be averted. In a screening promoted by the United States Center for Disease Control and Prevention, 148 different environmental chemicals were found in the blood and urine from the US population, indicating the extent of our exposure to environmental chemicals. Growing evidence suggests that environmental pollutants may cause diseases via epigenetic mechanism-regulated gene expression changes.[113]

The team also published a table listing the various diseases now linked by science to epigenetic effects – including leukaemia, breast, prostate and bowel cancers, schizophrenia, heart disease, traumatic brain damage, neurological disorders, Parkinson's disease, asthma and psoriasis.

Most of the studies that show epigenetic transfer of chemical damage from parent to child and grandchild have been conducted in laboratory animals, and generally at doses of the suspect chemical far higher than humans will normally encounter. Nevertheless, they prove that it can happen, and such proofs are multiplying as time goes by.

Epigenetic damage is not just confined to a few isolated cases but is becoming a serious public health issue warned Emma Marczylo and colleagues in a 2016 review. 'Identifying and understanding environmentally induced epigenetic change(s) that may lead to adverse outcomes is vital for protecting public health.' They urged the regular testing of chemicals for potential epigenetic effects, as well as other forms of poisoning and carcinogenesis, in order to develop sound prevention policies.[114]

Like some biblical curse, chemical poisoning and the harm it causes may thus not only affect the people immediately exposed but also the next generation . . . and the next . . . and the next.

Preventable Problem

Disease represents an awesome burden on humanity, our dreams and our ability to achieve them: for each person on the Planet, four months a year on average are lost to disability and premature death from disease.[115] Some of that is unavoidable: we all age and must die sometime. However, of the 165,000 people who die every day, about one in three dies before their time. This is known as the burden of preventable disease and much of it – at least half – is regarded by scientific medicine as readily preventable.

The major causes of death fall into two main categories – deaths from infectious diseases such as malaria, influenza, TB, pneumonia and Covid-19, and deaths from non-communicable diseases (NCDs) such as cancer, stroke, heart disease, diabetes, obesity and mental disorders (Figure 6.3). As we have seen, many

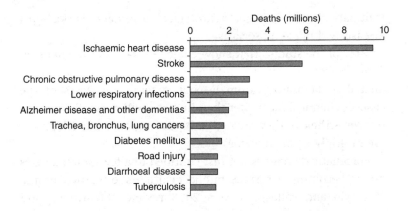

Figure 6.3. Top ten global causes of death. Source: WHO, 2018.

of the latter are now linked to the chemical environment in which the person lived.

Infectious diseases can be prevented by good public hygiene, vaccination and sound medical treatment. The NCDs must largely be prevented by removing the trigger for the disease from the person's living environment – whether it be an unhealthy diet, polluted air, dirty water or exposure to toxic vapours. For example, smokers who quit early gain up to ten extra years of life – and clean water, free of infectious organisms or pollutants, saves millions of lives.

There are two ways of looking at this issue – at the diseases caused by chemicals, and at the chemicals that are known to cause disease. For the sake of brevity, clarity and relevance to the reader, this book has taken the first approach. The second, what do certain chemicals actually *do* to you is taken by medical scientist Professor Alfred Poulos in his excellent book *The Secret Life of Chemicals*.[116] This deals with the threat according to chemical category – pesticides, plastics, heavy metals, indestructibles (such as dioxin and PBDE), workplace poisons, air pollutants, paper chemicals, fluorocarbons, radiation and so on, explaining the known health impacts of each and suggesting ways that we can reduce our exposure. Among these Professor Poulos lists: choosing safer foods, avoiding plastic packaging, filtering your drinking water (or storing rainwater), choosing a rural lifestyle, avoiding use of home and garden chemicals, using an indoor air purifier, and taking proper precautions in the workplace. However, he concludes, 'As both the diversity and quantity of pollutants increases ... it will indeed be surprising if there is no corresponding impact on our health and wellbeing. It is to be hoped that governments recognise that pollution is a problem with potentially catastrophic consequences for humankind and that action is taken before it is too late to reverse.'

This chapter has also made it clear that many new pandemics – worldwide diseases claiming millions of lives – are on the march, and that diseases once rare in the human population are becoming far more common. The evidence is also piling up in tens of thousands of scientific studies that many of these diseases relate to the man-made chemical flood that surrounds us.

Governments, by and large, accept the argument that people need clean water and air, safe food and smoke-free buildings. But to the larger chemical assault, described here, they remain stubbornly, inexplicably blind. Partly this is because the big picture is still fragmentary – there are tens of thousands of chemicals and they react in unpredictable ways with one another and with humans; providing the proof needed to tie a particular substance to a particular disease is typically a lot harder than proving that carbon dioxide causes the climate to change or that cigarettes cause cancer. Nevertheless, the evidence has now mounted to the point where a spate of new, otherwise inexplicable, pandemics makes it almost undeniable.

In short, poisoning by man-made chemicals has become the biggest preventable healthcare issue of our time – one that is on track to double and triple in coming decades.

The question for humanity is: do we wish to prevent it?

7 GETTING AWAY WITH MURDER

'The future depends on what you do today.'

Mahatma Gandhi

The bizarre epidemic at Minamata first came to medical attention in early 1956. It was reported, almost immediately, by a doctor employed at the chemical company's hospital, to the management and Board of the Chisso Corporation. He was formally instructed by the company to cease his inquiries into the source of the disease (although, courageously, he did not do so). By 1959 the local university had traced the culprit: organic mercury, one of many contaminants poured by the company into the bay on which the local fishing industry and families depended for their livelihoods and food. Angry fishermen stormed the factory offices and were met with blows and threats. Over the ensuing years of legal and physical battles some were bought off by the company with meagre compensation; others continued to fight.

But the pollution did not cease – and nor did the spreading sickness, with all its hideous symptoms and agonising deaths. For two decades, the company denied, lied and flung counter-accusations. It falsely claimed to have solved the problem. It bribed and threatened, set citizen against citizen, families against themselves, unionist against unionist. It hired lawyers and thugs and used them to terrorise protesting victims of the poisoning. It recruited sympathetic local and national government officials to its cause. It barricaded its headquarters with steel bars to keep angry protesters at bay. It met media inquiries with bluster and disinformation. It split the Japanese nation.

Eventually it palmed off the entire issue onto a government board of inquiry which returned equivocal findings. Finally, in 1973, Chisso ran up against an incorruptible judge who found the corporation guilty, saying:

> ... *a chemical plant, in discharging the waste water out of the plant incurs an obligation to be highly diligent; to confirm safety through researchers and studies regarding the presence of dangerous substances mixed in the waste water as well as their possible effects upon the animal, the plant, and the human body, always availing itself of the highest skill and knowledge; to provide necessary and maximum preventative measures ... in the final analysis ... no plant can be permitted to infringe on and run at the sacrifice of the lives and health of the regional residents.*

After a battle lasting nearly twenty years, the company president finally bowed in abject apology to the families his corporation had knowingly maimed and ravaged, with his forehead pressed to the ground in the traditional Japanese gesture of humble submission.[1]

In 1972 Eugene and Aileen Smith, who had followed and reported on the closing chapters of the human drama at Minamata, returned to the USA where, almost immediately, they were contacted by concerned citizens in Canada, recounting a tale of identical horror involving a chemical plant that was poisoning a local river, and the Canadian first people who depended on its fish. In 2014 the poisoned water crisis erupted in the town of Flint, Michigan. And so, on and on, the cycle goes.

Time and again, around the world, the pattern is repeated: a toxic release followed by corporate denial, misdirection, counterattack, media battles, protracted lawsuits (sometimes conclusive but usually delayed, costly and inconclusive, with much stalling); the complicity and corruption of government departments and officials, community anger, bitterness and suffering. Events surrounding the nuclear disaster at Fukushima, Japan, and revelations around the presence of chromium-6 in the drinking water supplies of two-thirds of US cities show that the ghost of Minamata has not been laid to rest.

Now, however, the contaminated site is not a suburb or a city. It is the world. Yet the response is no different.

Bureaucratic Intervention

The affair at Seveso, Italy, in 1976 involving a company, ICMESA, a subsidiary of Swiss industrial giant Hoffman LaRoche, highlights an ugly feature of such events – the complicity of governments in trying to silence community concern. One day after the accidental release of a poisonous gas during the production of herbicides the company voluntarily acknowledged the release had taken place. On day three, local health authorities announced there was 'no fear of any danger to the people living in areas surrounding the plant'. On day seven, the product (TCDD) was revealed to contain the carcinogen dioxin. On day twelve, the local government demurely stated: 'At this time there is no cloud of toxic gas,' and the following day said: 'Other health measures should not be considered necessary or urgent.' Meanwhile, the Regional Health Director proclaimed: 'Everything is under control'; however, on the same day the company's own medical director stated, 'the situation is very serious and drastic measures are called for'.

Caught in a lie, the local government initially tried to throw doubt on the medical director's credentials, then backflipped on the following day, announcing that 179 people should be evacuated. In all, 6 tonnes of deadly TCDD were dropped on an area of 18 square kilometres, exposing 37,000 people. Subsequent investigations revealed about 200 cases of serious chemical-induced skin inflammation and an unexplained increase in lung diseases, diabetes and certain cancers, notably breast cancer among younger women. Five employees – including two managers – of the chemical plant were charged, convicted, and then released on appeal without penalty. Twenty years later, in 1996, the event prompted the European Union to pass a law known as the

'Seveso Directive', requiring higher standards for the protection of the public.

Evading Responsibility

In India in the 1960s hunger was the focal issue. Ten of millions of people were starving, and the country's rulers saw a desperate need to build the world's largest democracy into a modern industrial state in order to create jobs for displaced people who were flooding into the cities from rural areas. In the regional city of Bhopal, in what was then Madhra Pradesh State, a solution to both was offered by Union Carbide (India) Limited (UCIL), an offshoot of the US Union Carbide company, which in the late 1960s began construction of a major chemical plant for the production of the pesticides that would be used to protect India's crops from attack by insects, thereby helping to relieve the food crisis. In the poorly planned city, whose population had doubled in size in less than twenty years, heavy industry jostled side-by-side with residential and commercial areas. Thousands of land-hungry squatters erected their homes within metres of the chemical plant perimeter.

In 1969 the plant was licensed to produce 5000 tonnes a year of a carbaryl-based insecticide, but a string of technical failures hampered development of the production process over the ensuing decade. By the time the unit was fully operational, demand from Indian farmers had moved in favour of cheaper, imported pesticides: production became uneconomic and was scaled back. The more money the plant lost, the more the quality of its operations deteriorated; skilled workers left and were replaced by inexperienced staff. In the crucial methyl isocyanate unit, the workforce was cut from the recommended level of three supervisors and twelve workers per shift to one supervisor and six staff. In December 1981, there was an ominous warning: a gas leak that led to fatalities.

In late 1984 India's then Prime Minister, Indira Ghandi, was assassinated. The news caused rioting to break out across Bhopal. making it even harder for employees of the plant to get to work. Meantime, a string of safety precautions began to break down around the two tanks that held the remaining 62 tonnes of deadly methyl isocyanate. Somehow, water from a washing operation got into pipes linked to the chemical storage tanks and backed up, causing pressure in the tanks to build dangerously over several hours. Soon after midnight, it reached a critical level – and erupted. Safety systems designed to trap or neutralise the agent failed and roughly 40 tonnes of deadly gas burst into the surrounding atmosphere. Alarms sounded. Workers rushed about attempting to fix things. Sleeping families in the surrounding suburbs awoke to the smell of gas, panicked and fled in all directions. Emergency services, broadly, failed to respond, but an Indian Army engineering unit – summoned by the manager of a neighbouring chemical plant – kept its head and began to organise an evacuation of the plant and the transfer of poisoned people to surrounding hospitals.

'Though there was defoliation of trees and some additional contamination of soil and lakes, the main impact of the accident was death and injury to humans and animals. Estimates of the number of immediate human deaths caused by the Bhopal gas cloud vary from the official Indian Government figure of approximately 2000 to the 10,000 favored by local activists,' wrote Massachusetts University's M. J. Peterson in his authoritative analysis of the disaster.[2] Another 200–300,000 were injured. Subsequent litigation established 3928 fatalities as due directly to the gas leak, but thousands more people were disabled for life and many widows and orphans suffered great hardship. In 2004, Amnesty International reported: 'The exact death toll on the night of the disaster will never be known. Estimates were made from burial grounds, from drivers of the fleet of 16 trucks moving bodies. 10,000 shrouds were made for Hindu victims alone; more than 7000 were burned on five funeral pyres. Many bodies were simply taken away by army trucks and dumped in mass graves or in the river Narmada.'[3]

The fight for fair compensation, justice, responsibility and the need to clean up the surrounding area and its contaminated groundwater dragged on for more than thirty-five years.

Activists claim a further 20,000 people have since died, with 120,000 suffering disability. Children are still being born with deformities attributed to the incident and the groundwater it contaminated. In all, some 574,000 people are thought to have been affected. The Indian Government, which originally sought compensation of $3.3 billion on behalf of the victims, settled for $470 million, which was finally paid over by 2003, much to the discontent of many victims.

No clean-up operation has ever taken place. 'It would be better if there was another gas leak which could kill us all and put us all out of this misery,' Omwati Yadav, 67, told the *Guardian* newspaper in 2019. She can still 'see the Union Carbide factory from the roof of her tiny one-room stone house, painted peppermint green with orange doors. Her body shaking with sobs, she cries out: "Thirty five years we have suffered through this, please just let it end. This is not life, this is not death, we are in the terrible place in between".'[4]

It was subsequently revealed that, besides the negligence of the Indian Government, the US Government also played a role in protecting Union Carbide, its managers and its subsequent owner, Dow Chemicals.[5]

Bhopal was the world's worst industrial disaster. It was also another case of the widespread complicity between government and corporate chemistry, at the cost of human lives. Comments medical scientist Alfred Poulos, 'the government's response has been ... we do not think that there is a problem but, if you do, prove it'.

Bitter Lessons

Bhopal, Seveso, Minamata, Flint, Hinkley and other famous cases have set a global pattern for chemical poisoning disasters.

It falls to the victims of poisoning to prove they have been injured by a chemical event and to demonstrate a clear link between cause and resulting injuries and deaths, pitted against the frequently combined efforts of corporation and government to hinder and discredit them. Rarely is the suspect factory or industry required to demonstrate that its processes and products were safe, or that it upheld high standards of public safety at the time of the event. Seldom are its management or owners brought to justice. Furthermore, while compensation and apologies may ultimately be extracted after years, almost never do they make good the lives laid waste.

The significance of these classic 'chemical battles' between large and powerful corporations and small groups of affected citizens is that they have established an unproductive pattern for conflict and confrontation. In this, the company typically digs in, denies responsibility, often falsifies scientific findings and fights through the courts, with tacit or explicit support from government. This response provokes anger and hostility from the public, environmentalists and the media; consequently, the scope for resolving such issues rationally and without bitterness is much diminished. Fearful of public outrage or loss in 'shareholder value', industry seals itself behind lawyers, spin doctors, tame scientists and razor wire, while governments seek to muddy the waters for their own protection. Fearful of such conflicts, even well-run companies and industries take refuge in secrecy or flee offshore.

However, these events constitute mere pixels in the image of the global chemical avalanche. They are generally local, providing no model for dealing with chemicals that have disseminated worldwide in the Earth system, causing harm to hundreds of millions across many countries, as described in the previous chapter. They are mere foreshocks of an earthquake yet to come.

There is one chemical with the capacity to harm the future of every person on the Planet: carbon dioxide. Its excessive release by human activity is scientifically linked, beyond doubt, to the heating of the Planet and the perturbation of its climatic and weather systems. Yet the evidence is frequently denied, disputed, diverted and dismissed by the industry most responsible: fossil fuels. The climate crisis is a vivid illustration of how 250 giant oil, coal and gas corporations – responsible for the lion's share (71%) of global greenhouse emissions[6] – have combined to delay, dilute and defeat attempts by governments and

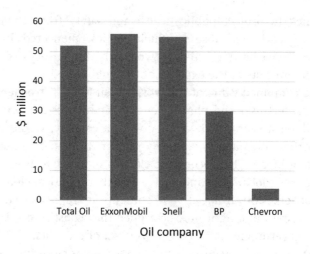

Figure 7.1. Spending by oil majors on 'climate branding' in $m, 2018. Source: InfluenceMap.

concerned citizens to head off a climate crisis by reducing fossil fuel use.

According to UK-based analysts InfluenceMap, in the three years following the 2015 Paris Climate Agreement, the five largest publicly traded oil and gas majors (ExxonMobil, Royal Dutch Shell, Chevron, BP and Total) invested more than $1 billion of their shareholder funds in climate misinformation and counter-lobbying (Figure 7.1).[7] A separate study found five oil companies had spent $3.6 billion over three decades on advertising designed to influence the artificially manufactured 'climate debate'.[8]

The world oil industry and the petrochemical industry are joined at the hip. The vast bulk of the world's manufactured chemicals, plastics, fertilisers and pharmaceuticals are made from petroleum, gas and coal. As renewable energy and electric vehicles take over the market, the fossil fuels sector is desperately seeking alternative outlets for its product.

Consequently it is no great surprise that the same tactics employed by oil majors to disrupt and sabotage safe climate policy are also to be found in the actions of the petrochemical sector to distract, divert, deceive and block laws and regulations designed to protect public health and safety from chemical-related death and disease. Such tactics, pioneered by Big Tobacco, have been honed over two-thirds of a century to defend private interests at the expense of the public good.

The combined value of the world fossil fuels, petrochemical, plastics, fertiliser and pharmaceuticals industries is more than $6 trillion a year which, if they were a country, would make them the third largest economy on Earth, behind China and ahead of Japan. Their economic clout is more than three times that of the global arms industry – and their political clout is in proportion. Most, if not all, governments are highly sensitive to their wishes. Indeed, it may be argued they constitute a *de facto* world government, for a very narrow set of interests.

Just as global warming is now a struggle for the future survival of the world's children and grandchildren on a habitable Earth, so too the even larger chemical avalanche constitutes the battleground for their health and safety on an increasingly poisoned Planet. A battle that the public is losing. 'Are we completely powerless in the face of what appears to be a "Chemical Armageddon"?' wonders Professor Alfred Poulos.[9]

Dirty Dozen

Of the tens of thousands of toxic and carcinogenic substances that have been developed and released globally in the past century, only a handful have ever been banned. The Stockholm Convention on Persistent Organic Pollutants was set up in 2001 by the United Nations to eliminate chemicals that persist in the food chain and environment: the Convention has 179 parties, including the European Union, which means that most countries in the world support it, at least verbally.

In 2004, the Convention outlawed nine of a set of twelve chemicals, dubbed the 'dirty dozen', and called for the remaining three to be restricted. It subsequently banned a further nine substances in 2010, adding up to a grand total of twenty-one substances over a period of two decades. Even so, the bans only have force in countries that signed and ratified the convention.[10]

At such a rate of progress – one chemical per year – it will take a third of a million years to test and ban, restrict or approve the full suite of registered man-made chemicals. This is nine times longer than *Homo sapiens sapiens* has actually existed on Earth. Even then such substances will only be banned in certain countries, which will still be powerless to prevent the pollutants entering their borders or their citizens via the Anthropogenic Chemical Circulation and the six global pathways described in Chapter 2. This estimation of progress takes no account of the thousands of new or repurposed chemicals being released every year.

The picture is replicated at national level. The US EPA, for example, has so far banned thirteen of the 84,000 chemicals it says are used in that country. Furthermore, the USA annually releases more than '1.5 million metric tonnes of chemicals that are persistent, bioaccumulative, and toxic (PBTs); over 756,000 metric tonnes of known or suspected carcinogens; and nearly 667,000 metric tonnes of chemicals that are considered reproductive or developmental toxicants', according to the UN Environment Programme (UNEP).[11]

Europe takes a tougher stance: in cosmetics, for example, the EC has banned 1328 substances from use in cosmetics, whereas America has banned eleven. More significantly, Europe requires chemical manufacturers to prove their products are safe before they can be legally marketed. As the UK *Guardian* newspaper summed it up: 'The clout of powerful industry interests, combined with a regulatory system that demands a high level of proof of harm before any action is taken, has led to the American public being routinely exposed to chemicals that have been rubbed out of the lives of people in countries such as the UK, Germany and France.'[12]

Regrettably, the widely used phrase 'dirty dozen' has created an impression in the minds of many citizens that the number of toxic chemicals is very small, and hence is fairly easily reined in.

In reality, UNEP estimates, around 30 per cent of all man-made chemicals released are toxic in one way or another. The European Environment Agency stated that 62 per cent of the 345 million tonnes of chemicals consumed in the EU in 2016 were hazardous to health.[13] On the basis of such estimates, from 750 million to 1.5 billion tonnes of chemicals manufactured each year are dangerous to humans – and this output is set to double by the 2030s. To this colossal burden of poison must be added the far larger tsunami of unintentional pollutants emitted by mining, manufacturing, agriculture, energy production and other activities. For these, as the examples above demonstrate, there is almost no hope of cleaning up the human living environment with regulation alone.

Torn Safety Nets

While global chemical use is forecast to intensify, growing by around 3 per cent per year up to 2050, the world's ability to regulate and restrict it is weakening.

The main reason for this is that, in their efforts to evade regulators, chemical corporations are winding back their operations in the developed world and moving to more poorly regulated countries, mainly in Asia. In the first two decades of this century, chemical output in Asian countries grew three to five times faster than in North America and Europe.

In the beginning, big chemical companies saw the developing world as a promising new market for chemical products that were difficult to sell, or in some cases were banned, in developed countries. It wasn't long before these companies realised that the lack of regulation, supervision and enforcement in developing nations, as well as their low wages and lack of workforce health and safety measures, made them highly attractive as places to manufacture toxic chemicals or carry out polluting activities. Regulation in the newly industrialising world is often rudimentary, enforcement lax and corruption of officialdom is rife. By choosing such places to relocate to, the chemical industry is deliberately placing itself beyond the effective reach of the law in well-regulated countries as well as exposing citizens and consumers in the developing world to a rapidly growing toxic load. And that is just part of the problem.

As developing countries and those in economic transition increase their economic production, related chemical releases have raised concerns over adverse human and environmental effects. Chemical contamination and waste associated with industrial sectors of importance in developing countries include pesticides from agricultural runoff; heavy metals associated with cement production; dioxin associated with electronics recycling; mercury and other heavy metals associated with mining and coal combustion; butyl tins, heavy metals, and asbestos released during ship breaking; heavy metals associated with tanneries; mutagenic dyes, heavy metals and other pollutants associated with textile production; and toxic metals, solvents, polymers, and flame retardants used in electronics manufacturing,

says UNEP.[14]

A vivid example of how this plays out was the discovery, in Australia, of the banned chemical dioxin in weedkiller on sale in home garden centres. Dioxin, it may be recalled, is a highly toxic byproduct of pesticide manufacturing that was the chief substance of concern in the Agent Orange scandal which ruined so many lives of Vietnamese villagers and US and Australian servicemen during the Vietnam War. Banned in 2004 under the Stockholm Convention, many people hoped it was gone for good. However, when the Western chemicals industry relocated to China, it brought with it many of the dirty, dangerous old manufacturing systems (like those used at Bhopal). As a result, cut-price herbicides made in China still contained deadly dioxin – and this was found on sale to home gardeners in Melbourne in 2013.[15]

China has sold numerous toxic products to consumers worldwide, including everything from children's toys to adult vitamins to pet food, furniture and even fresh vegetables. The US government regularly blocks more poisonous or faulty products at the border from China than from any other nation. In March

2020, for example, the Food and Drug Administration issued 224 import refusal notices for Chinese goods, compared to 113 for India and 214 for Mexico, the two next most prolific purveyors of dodgy merchandise.[16] However, all these safety nets have big holes – and officialdom has no way of blocking the Anthropogenic Chemical Circulation or the Six Pathways.

At the same time, safety protections for citizens' lives and health are being rolled back. The trend began in the USA with the deliberate efforts of the Trump Administration to dismantle 125 environmental laws which it claimed were hindering industry, while giving free rein to fossil fuels and other polluters.[17] Two Harvard scientists calculated that just sixty of these changes would cost the USA around 80,000 additional lives, and 1.5 million cases of disease, per decade.[18] In most things chemical the USA is a world trendsetter, and the danger is that other countries will follow them into increased citizen carnage. Furthermore, rolling back safety laws at the behest of the chemical and fossil fuels industry in one country encourages them to believe they can do it everywhere.

During the 2020s, the newly industrialising world will become the main seat of global chemical manufacturing power, with China the fastest growing and largest producer and already a research leader.[19] While environmental laws are growing stronger in China, they still lag well behind the developed world.

Even if the will existed in every nation to curb the tsunami of new and untested substances or old and known toxics being unleashed on humanity, the means to do it will not exist – at least for the foreseeable future. Other ways must be found to stem the flood.

Seeking Standards

Alarmed at mounting public fury over chemical incidents and the rapidly accumulating scientific evidence of chronic harms, especially to children, as well as tougher rules in places like the EC, professional chemistry bodies in well-regulated countries began adopting voluntary standards.

The industry's international *Responsible Care®* initiative started in Canada in 1984, was adopted in the USA in 1988, and by

2020 had spread to sixty-eight of the world's countries. (That still left about two-thirds of countries in the irresponsible category.) The industry states Responsible Care is 'the chemical industry's world-class environmental, health, safety and security performance initiative. It's our commitment to doing more, and doing better. Responsible Care companies are industry leaders, playing a vital part to ensure that the business of chemistry is safe, secure and sustainable.'[20]

However, closer inspection of the claimed results for Responsible Care reveals a far higher focus on industrial health and safety than on societal health and safety. Worker safety is extremely important and the industry's published figure of a 66 per cent reduction in work-related injuries and illness is laudable. However, there are no metrics to suggest the chemical industry is doing much to curb the 9 million annual death toll attributed by world health authorities and medical science to chemical poisoning in all its manifold forms – or is even much bothered by it.

The twelve principles that guide Responsible Care are set out on the American Chemical Association's website. Most of them focus on the industry and its needs. Only three invoke the wider issues of public and environmental safety. None of them acknowledge the elephantine questions of total chemical emissions, mixtures, global poisoning and its fatalities or the wildlife extinction it is causing.[21]

Britain's Royal Society of Chemistry operates a somewhat ambivalent Code of Conduct for its members, in which it says, among other statements:

All members have responsibilities arising from their duty to serve the public interest, and should be concerned with the progress of the chemical sciences. The RSC does not condone any attempt to coerce its members into refraining from lawful activity. It expects its members to use their professional skills to:

- *Advance the welfare of society, particularly in the fields of health, safety and the environment.*

- *Advocate suitable precautions against possible harmful side-effects of science and technology.*
- *Identify the risks of scientific activities, and take an active interest in safety throughout their organisations.*
- *Undertake any lawful scientific activity as required even if in an area which arouses adverse publicity.*
- *Use their knowledge and experience for the protection and improvement of the environment.*[22]

In the fourth point, the RSC evidently expects its member chemists to do what the boss tells them, so long as it is 'lawful', even if they or the public are concerned over a possible chemical impact affecting the welfare of society. This is clearly a position the RSC needs to rethink from an ethical standpoint.

Australia's Royal Australian Chemical Institute (RACI), by contrast, stated clearly in its Code of Ethics where the first duty of a chemist lies: 'a member shall endeavour to advance the honour, integrity and dignity of the profession of chemistry. However, notwithstanding this or any other By-law, the responsibility for the welfare, health and safety of the community shall at all times take precedence . . .'[23]

While codes of conduct appear superficially to betoken good intentions on the part of the chemical industry, they also act as a smokescreen for its covert activities in seducing governments to weaken their rules and commissioning pliant scientists to dispute or undermine public good research warning of chemical harms.

The US Environmental Protection Agency, for example, states unambiguously that its mission is 'to protect human health and the environment', yet under the pro-pollution Trump regime it was forced to cancel rules protecting air quality and water safety, toxic chemicals, mineral and energy extraction, and healthy cities and to go silent on the impacts of climate pollution. This demonstrates how easy it is to dismantle public health and safety and environmental protections – and how flimsy, in reality, is the regulatory shield that guards humanity.

Such codes and guidelines, it should be noted, are observed only by legitimate, ethically run chemical enterprises and professional organisations in well-regulated countries – they do not govern the behaviour of chemical firms in the other two-thirds of the world, nor can they curtail the activities and behaviour of criminals, chemical weapons makers, nor miners and mineral processors, the electronics sector, the food industry or the energy, transport and building sectors. In short, self-regulatory codes and guidelines apply only to a fraction of the sources of the human-emitted chemical avalanche to which we are all exposed.

Regrettably, even in this restricted set of cases, the pattern of global development revealed by UNEP indicates that large parts of the world's most polluting industries are relocating away from countries where high standards of regulation and compliance – and high costs – apply. Many companies willing to observe such standards in their home countries are hedging their bets by operating subsidiaries in more loosely governed or readily corrupted places. At the same time it is evident that polluting industries maintain subtle but unrelenting political and economic pressure on governments and agencies in their homelands to undermine both regulation and scrutiny, distort the science and streamline the approval of new substances or the re-use of old substances without adequate testing.[24]

Current international regulation of dangerous chemicals may be likened to trying to control traffic in a big city one car at a time, instead of imposing road rules and traffic lights to be observed by all. Decades ago, when the global number of chemicals was small and their dissemination mainly local, having regulations specific to certain substances in particular countries may have worked; in today's globalised economy and with the dispersal of tens of thousands of man-made substances throughout the Earth system, it has no chance of success.

'There is increasing recognition among governments, non-governmental organizations and the public that human health and the environment are being compromised by the current arrangements for managing chemicals and hazardous wastes,' commented UNEP. 'These concerns take on a new level of urgency as the quantity and range of new and existing chemicals grow rapidly in developing countries and economies in transition.'[25]

When Rules Fail

Man-made chemicals are so widespread in the world today because they are very useful, very valuable, very profitable and help to enhance billions of lives. They are a central enabling technology in the modern global economy. They are never going to be universally banned – and nor should they be.

But neither should they flood the Earth uncontrolled.

The magnitude of our chemical – especially toxic chemical – exposure has crept up on the human population unawares. Even the chemicals industry itself, with its self-focus on specific substances and products, appears to have little grasp of the current Planet-wide and pandemic impact of its products and their interactions with substances from other industries and with all life forms; or else chooses to avert its eyes. Other polluting industries, such as mining, food and energy production, construction and transport, also make insufficient effort to understand and curb the Planetary impacts of the pollution they cause.

This tendency to view the Earth in terms of one's own locality and business interests may have been appropriate a hundred years ago – but in the mid-twenty-first century, with the human population soaring to 10 or 11 billion and a redoubling in demand for all material goods, such complacency – and feigned ignorance – is now a mortal threat.

Continued and tougher regulation of chemicals by national governments is essential, let that be clear. So are industry codes and self-regulatory standards, provided they recognise the issue of global emissions and the need to correct it. But the overwhelming evidence to date says that neither regulations nor codes can be relied on to curb the chemical avalanche, the poisoning of the Planet and human species. Put simply:

- Regulation has so far banned fewer than 1 per cent of all intentionally made dangerous chemicals – and then only in certain countries. Regulation has not prevented their legal production in others, illicit production and use, or their circulation and persistence in the environment (though it may have restricted it in some jurisdictions for a limited number of substances).

- Regulation has little prospect of controlling the production and use of tens of thousands of substances whose health thresholds are still either unknown or poorly defined.
- In all likelihood, regulation will struggle to deal effectively with the problem of chemical mixtures (as innocuous substances may have toxic effects in combination) especially in the diet, home, workplace and city environment. Also, making rules cannot prevent man-made chemicals themselves from forming adventitious new compounds as they interact with other substances, both man-made and natural, or degrade.
- Regulation cannot stop emitting industries transferring their base of production from well-regulated countries to weakly or unregulated countries.
- Regulation has so far been unable to enforce the thorough, objective health testing of each new chemical and its byproducts prior to commercial release.
- Regulation may limit, but cannot prevent, epigenetic damage to future human generations inflicted by exposure of people in the present generation.
- The restriction or banning of chemicals one or several at a time will, at current rates, take millennia, besides incurring great expense. Past experience suggests that many industries will fight tooth and nail to protect their profits and stall public safety measures.
- Regulation has not succeeded in banning – or even limiting – the growth of illegal chemical production and contamination of the environment, water and food chain by organised crime worldwide. Indeed, production of recreational drugs and 'performance enhancers' appears to be increasing. Furthermore, regulation has so far not even prevented the contamination of the environment by legal medical drugs.

In 2002, the Earth Summit adopted the 'Johannesburg Plan', which had at its heart a vision that, by 2020, chemicals will be 'used and produced in ways that lead to the minimization of significant adverse effects on human health and the environment'.

By 2020, it was abundantly clear the plan had failed and chemical emissions were soaring. The UNEP Global Chemicals Outlook of 2019 conceded as much.

National and international regulation *alone* cannot restrain man-made chemical output or the creation of new substances with unknown impacts. It is evident that it cannot even 'minimise' its effects, whatever that may mean. At best, such goals serve to keep the issue in the public mind, encourage stricter controls in the best-run countries and set standards which others may eventually follow.

Who Is Responsible?

This chapter seeks to underline that while regulation and industry codes of conduct are essential, laudable and should go much further, on their own they cannot restrain the chemical behemoth that has been unleashed by a globalised economy, nor can they prevent the beast from doubling or tripling in size by mid-century. Neither can long, legal battles between corporations and their victims prevent future chemical catastrophes, though they may eventually lead to stronger local rules.

It is also time for critics of the chemical industry to recognise that blaming industry over individual events and demanding tougher laws will not solve the problem either. Indeed, blaming industry is likely only to harden its resolve to become less transparent and fight fiercer rear-guard actions – and will simply encourage manufacturers to move to less strictly regulated places where the government is more pliable or sympathetic.[26] It is far, far better that a contaminating industry be based in countries where its own standards – and government standards – are high, rather than move to a place where it will escape scrutiny yet continue to compound global contamination.

It is time for everyone to realise that we ourselves, by virtue of the demands we place on industry for ever higher living standards and cheaper products, are complicit in the poisoning of the

Earth system by human chemical activity. It is our rising demand for material goods, our swollen numbers and our obsession with 'growth' that has unleashed the chemical avalanche.

Such poisoning is an inevitable consequence of our own wishes, wants, needs, whims and fashions expressed in the consumer society, which the market is trying to satisfy, often by chemical means. It is our desire for comfort, convenience, safety – and, above all, for cheaper products – that drives the market and creates the monetary incentives for chemicals to be made or emitted.

Mahatma Ghandi once said, 'The future depends on what you do today': we need to appreciate that our demands come at a high price, and that price is increasingly personal. If we use toxins against insects or weeds, or contaminate our homes, groundwater and soils, consume food produced with chemicals, cosmetics, drugs, electronics and manufactured goods, if we use fossil fuels, minerals and electricity, then we unavoidably end up contaminating ourselves, our children and everyone else far into the future.

On a crowded Planet, every act of consumption has chemical consequences.

Some of these consequences are lethal. It is time we understand this, at the level of the species as well as the individual. It is a matter of personal, as well as collective, responsibility.

Human chemical emissions are directly responsible for killing around 9 million people per year and disabling 86 million. Indirectly they afflict tens of millions more with a burden of disease, lost work and caring for its victims, that grows with each passing year. The scientific evidence is amassing that our chemical emissions harm everybody, everywhere; that they impair the habitability of the Earth itself for humans and life in general.

We are all participants and shareholders in this process, though we may not care to admit it or face up to the responsibility it bears. Yet, in our hearts, everyone who reflects on this will know it to be true.

In a sense, we are all getting away with murder.

8 CLEAN UP SOCIETY

'A pandemic of neurodevelopmental toxicity caused by industrial chemicals is, in theory, preventable.'[1]

Philippe Grandjean and Philip Landrigan

'And the Winner Is ...'

On 23 September 1993, thousands of Australians jumped for joy as Olympic Games boss Juan Antonio Samaranch proclaimed, '... and the winner is ... Sydney'. Unbeknownst to the hundreds of thousands of athletes and visitors who flocked to the Sydney Olympics in 2000, the words also triggered one of the largest and most successful environmental clean-up efforts on record.[2]

Beneath what became the Olympic site and village, quietly festering away, there lay an estimated 9 million cubic metres of waste and contaminated soil spread over 400 hectares of the 760 hectare site. A toxic outgrowth of the 'consumer society', this garbage had accumulated rapidly since the 1950s and included petroleum waste, unexploded ordnance from an old military store, acid sulphate soils, illegally dumped wastes along the waterways (including persistent organic pollutants, polycyclic aromatic hydrocarbons etc.), dredged sediments, municipal waste in managed tips, industrial waste (such as rubble, power station fly ash, gasworks waste and asbestos) and contamination from burning pits, chemical leaks and chemical use. In terms of what it was jettisoning from its memory, Sydney was no different from any other big industrial city around the Planet. All metropolises have hideously toxic things underfoot, about which we prefer to be ignorant – but which long remain a sleeping, seeping menace.

The clean-up of this industrial filth became the largest project of its kind in Australia and remains one of the most significant positive environmental legacies of the Sydney 2000 Olympic and Paralympic Games. The site was extensively investigated using exploratory boreholes and a remediation plan adopted which involved safely containing and, where possible, treating waste on site, rather than relocating it to other places. Nine million cubic metres of waste were recovered, consolidated and relocated to designated waste containment mounds. These were capped, sealed, landscaped and turned into 160 hectares of parkland. Collection and transfer systems were built to prevent chemicals leaching into the environment. Four hundred tonnes of soil contaminated with hydrocarbons and hazardous chemical waste were treated in a two-stage thermal desorption and destruction process. Dust, vapours, noise and water were all continually monitored.[3] Nearby Homebush Bay became, pre-Olympics, the target of a major clean-up effort. Previously it was regarded as 'one of the most contaminated waterways in the world, due to the cocktail of organic contaminants', including DDT, dioxins and furans escaping from a former Union Carbide plant,[4] and was so heavily polluted that authorities banned residents from eating the local fish.

When it was completed, the Sydney Olympic Park was about as clean, safe and healthy as the technology of the day could render it. Besides proving it possible to make safe some of the world's most horribly polluted land, a major benefit was the dramatic gain in local land values – from being next to worthless, the remediated land rapidly acquired inner city property values and attracted contracts for major housing, recreational and retail developments. The clean-up operation cost AU$137 million, but the renovated land ultimately sold for around AU$600 million.[5] Not many investments generate such an attractive rate of return so quickly. This illustrates the economic gains to be had from cleaning up yesterday's (and today's) urban messes, not to mention the long-term costs of healthcare and premature death that the clean-up may have averted. The enduring message from the 2000 Olympics is quite simply that a cleaner world is a more prosperous and profitable world – as well as a safer one.

The case of the Sydney Olympics clean-up shows it is feasible to render a badly contaminated site reasonably clean and safe using today's advanced remediation methods. Other celebrated clean-ups are described below.

Safer Singapore

The island state of Singapore has a commitment to cleaning up its environment dating back to a tree-planting campaign in the 1960s and the cleansing of the once famously foul Singapore River in the 1980s.[6] Today, a major challenge is converting Singapore's 7 million tonnes of garbage a year to zero waste; the State faces a deadline to end waste because its only major landfill is due to run out of space in 2035.[7]

Singapore is one of the most progressive nations on Earth in rethinking its patterns of industrial and citizen behaviour: 'The "take, make and dispose" way of consumption – also known as the linear economy model – is no longer sustainable,' says its government.[8] Furthermore, unlike many places, Singapore recognises that climate change is a direct threat to human survival and must be met by urgent decarbonisation. Being a small island state, Singapore is also acutely conscious of the looming problem of resource scarcity and is applying the full power of science to solving it. Awareness of all these issues is driving a national impetus to clean up, recycle and end pollution and food waste.

The cornerstone of the State's waste management is *Clean and Green Singapore*, a campaign to educate and engage the whole community, from children up, in recycling and sustainable behaviour.[9] It is a model for how citizens must behave if they wish to lead healthy lives in an increasingly toxic world.

A history of air pollution problems – many blown in on the wind from other Asian countries – has bred in Singaporeans an acute awareness of chemical pollution of air, water, food and the living environment, and they were more than willing to comply.

However, with 2000 companies using toxic chemicals in Singapore, chemicals are less tightly regulated than, say, in the European Community (EC). The National Environment Agency explains, 'Singapore is an island of about 620 square kilometres in area. The average population density in Singapore is about 4,000 people per square kilometre. Such high population densities make it imperative for hazardous substances to be controlled so that public exposure to accidental release is, if not avoided, minimised.' However the focus remains chiefly on air quality and on chemical disaster prevention, rather than the diffuse chemical avalanche engulfing all aspects of daily life in Asia.

Super Clean-Up

Since the US Environmental Protection Agency's Superfund clean-up scheme was set up in 1980, it has located and analysed more than 91,000 potentially hazardous waste sites and implemented a programme of remediation and cost recovery. By 2020 some 1400 sites were on its priority treatment list and over 400 had been cleaned up in what had become a world model for dealing with dangerous local cases of chemical pollution. The clean-ups were funded by the taxpayer, but with substantial industry contributions under the 'polluter pays' principle, in cases where the polluter was still in business. The US EPA publishes an extensive list of treated sites.[10] A pleasing feature of Superfund clean-ups is the close consultation with affected communities, which involves being open about the problem, listening to their concerns, and learning what they want done about the issue.

The EC has a strong focus on the clean-up of dirty old industrial sites – but its results are much spottier, owing to national differences in approach to the problem. Since 2015 it has operated a network of 1500 contamination experts in thirty-nine countries to try to highlight best-practice approaches to clean-up across the Community. From this it estimates there are 3.3 million polluted industrial sites across the EC in need of cleansing.[11]

Owing to its explosive industrial development in recent years China is heavily affected with chemical contamination. Its efforts to come to grips with the problem overall are fragmented by the different approaches adopted by provincial governments, although most of them are taking some action to reduce pollution. Chinese scientists consider the PRC 'still lacks effective strategies on issues such as remediation standards, funding resources, accountability, and stakeholders coordination'.[12] A national survey found that one-fifth of China's farmland – and hence food chain – was heavily contaminated and would cost an almost unaffordable $1.3 trillion to clean up.[13] The need for extensive remediation measures was underlined by the massive Tianjin chemical explosion in 2015, which killed 114 people, injured more than 700, left 69 still missing and damaged 17,000 homes. The event prompted a huge effort to cleanse the waters of the local river, groundwater and soils.[14]

The experience around the world is that contaminated industrial sites can, with sufficient time, money and skill, be cleansed to a point where they are safe for people to live and work there – and that it can even be profitable to do so. However, these sites are mere pinpricks in the map of global contamination – and their repair contributes little to the prevention of future global contamination or to limit the Anthropogenic Chemical Circulation. While it is essential to address local toxicity, it is irrational to imagine that cleaning up single sites will ever do much to arrest or diminish the chemical avalanche in its entirety.

Cleaning single sites does nothing to detoxify a chemically fouled world, each of whose inhabitants is, in ourselves, a contaminated site.

To heal our world, prevention is the only cure.

Going Global

Compared to the Olympic Park clean-up operation, the challenge of reducing the individual and Planet-wide burden of toxic man-made chemicals is very large indeed: it involves preventing or controlling a fresh influx of 120–220 billion tonnes of human-emitted substances each single year. It involves objectively analysing which of our present and future chemical releases are dangerous and which may be relatively harmless.

This may at first glance appear to be an insurmountable task. Unlike climate change, which is triggered by a relatively small group of chemicals from a narrow range of activities, we are here talking about a vast array of emissions from every conceivable human activity from mining to lipstick making. However, with global awareness, good technology, partnership and a clear desire to care for ourselves, our children and our Earth, it may be possible for humanity to overcome it.

There is little that humans have done that cannot in some way be improved, rectified, replaced or undone. The challenge is to quickly build a worldwide consensus for rapid and concerted action against future poisoning that makes all these things happen.

The first point for all citizens to understand, however, is that Earth system toxification will not be curbed, solved, or even minimised if left to government regulation and industry compliance alone – even when they act from the best of intentions – so long as billions of consumers all around the world persist in issuing economic commands for the mass release of toxins. And while it is a global problem, it cannot be solved by enforcement at a global level, because international agencies have little or no power to act at national, local and industry levels.

However, global agencies do have the ability to influence, guide, inform, educate and provide templates for regional, continental or Planet-wide action. Because, like it or not, we citizens of the world are all in this together: we are drowning in our own filth. Even if individual countries, such as those of Europe, are willing to curb their own toxic releases, they and their citizens cannot escape the worldwide flux of contaminants in air, water, food, wildlife, imported products and damaged human genes.

There is no clean country in a dirty world.

Eugene Smith recorded his searing images of Minamata because he hoped that, through public awareness, such evidence might provide an indelible lesson for society, to forever guard against any recurrence. Yet, for all the rise in public awareness, cases of chemical poisoning have happened time and again and are still happening, on an ever-increasing and now ubiquitous scale. Every indicator now points to the poisoning continuing to escalate for the remainder of human history, until our health, intelligence, living environment and wellbeing are damaged beyond repair.

Unless something profound changes.

Before we address what that change entails, let us review briefly the many methods currently available to clean up our communities, our industries, our countries and the Planet, and the main instruments that make it possible.

Clean Rules

Many countries around the world now have well-established regulatory structures and expert environmental agencies for keeping track of certain chemicals in their various forms – as commodities, constituents of products, environmental pollutants, occupational and public health hazards and wastes – and regulating their use. These agencies share chemical knowledge among themselves, making the task of the identification of toxins less onerous and costly than it would be for each individual body or nation. They also set standards of excellence which others can follow. They perform an invaluable educative role, but one that tends to be restricted to the thinking public; their publications are seldom easy for people without science training to understand.

However, these agencies are also nervous about antagonising industry and are vulnerable to politicians who want to undo their good work. Environmental protection agencies have also been unable to keep pace with the tide of new chemical releases, safety testing of these new chemicals, and the issue of how to deal with chronic low-level poisoning of their populations from a multitude of chemical mixtures originating from numerous sources. Such agencies also find themselves under unrelenting industry attack for creating 'needless barriers' to business growth. This often makes them loath to act on a precautionary basis, as citizens might wish and wisdom might dictate.

As explained in the preceding chapter, in some countries, the chemical, food and other manufacturing industries have self-imposed safety and health standards and ethical guidelines. These tend to apply to particular products and processes, and not to the combined chemical burden of the human population or to the mixtures of poisons in our food, air, water, homes and cities. They ignore the forest of contamination for a handful of single toxic trees. Furthermore, such codes of practice are less common and even less effective in industries such as mining, mineral processing, energy production, transport, construction, agriculture and so on – all of which are mass emitters of chemicals into the global environment.

The United Nations' Strategic Approach to International Chemicals Management (SAICM) and the various chemical-related multilateral agreements (see Table 8.1) provide a range of voluntary and legally binding frameworks designed to encourage better management of chemicals. SAICM was formed in 2006 to give impetus to the Johannesburg Declaration on Sustainable Development of 2002, which declared that by 2020 chemicals should be 'produced and used in ways that minimise significant adverse impacts on human health and the environment'.

However, these international instruments operate chiefly by suasion and have little real power to ban chemicals or control their use. That is left to government agencies in individual countries, which are often under industry pressure to do as little as possible. Also, these global agreements tend mainly to focus on individual substances and pathways that have been shown to cause harm, rather than on the central issue of rising chronic exposure and the combined chemical burden of the entire human population.

Table 8.1 Key international chemical treaties and agreements

- Convention on Long-Range Transboundary Air Pollution (1983)
- Montreal Protocol on Substances that Deplete the Ozone Layer (1989)
- Basel Convention on the Control of Transboundary Movements of
  Hazardous Wastes and Their Disposal (1992)
- International Convention on Oil Pollution (1995)
- International Labour Organization (ILO) Chemicals Convention (1993) and
  Prevention of Major Industrial Accidents Convention (1997)
- World Summit on Sustainable Development, Johannesburg (2002)
- Rotterdam Convention on the Prior Informed Consent Procedure for
  Certain Hazardous Chemicals and Pesticides in International Trade (2004)
- Stockholm Convention on Persistent Organic Pollutants (POPs) (2004)
- Dubai Declaration and Strategic Approach to International Chemicals
  Management (2006)
- WHO International Health Regulations (2007)
- World Business Council for Sustainable Development (WBCSD) Chemical
  Sector SDG Roadmap (2017)
- Minamata Convention on Mercury (2017)

Source: UN agencies, 2020.

As with climate action, world progress to check the chemical avalanche has been glacial, contested every step of the way by industry which professes to care – but self-evidently does not.

The Sustainable Development Goals embody other international goals to check chemical impacts, so far without much success:

- SDG 12.4 – 'by 2020, achieve the environmentally sound management of chemicals and all wastes throughout their life cycle . . .' has already failed.
- SDG 3.9 – 'by 2030 substantially reduce the number of deaths and illnesses from hazardous chemicals and air, water, and soil pollution and contamination' has not curbed the mounting death toll, and in any case applies chiefly to air pollution, not the chemical assault in general.

- SDG 6.9 – 'by 2030, improve water quality by reducing pollution, eliminating dumping and minimizing release of hazardous chemicals ...' has equally poor prospects, and is limited to water.

In short, UNEP asserts, 'Despite global agreement reached at high-level UN conferences and significant action already taken, scientists continue to express concerns regarding the lack of progress towards the sound management of chemicals and waste.'[15]

UNEP and others advocate a 'multi-stakeholder' approach to dealing with chemicals which brings together government, industry and civil society in an effort to reduce the human chemical burden. However, while industry and government are sometimes in alignment, the voices of citizens, consumers and chemical victims are seldom heard, while the interests of children – the future citizens of the Planet – are unrepresented. Building even a national consensus for action on chemicals is painfully slow – and in the meantime there is no escape for anyone from the rising tide of contamination reaching us from places where such a consensus may never be attained.

In its 2019 Global Chemicals Outlook (GCO) II report, UNEP mapped a path forward at global level towards Planetary clean-up:

1. Develop effective management systems, including strengthening national chemical laws and policies;
2. Mobilize more resources for effective legislation, implementation and enforcement, particularly in developing countries and economies in transition;
3. Assess and communicate hazards: fill global data and knowledge gaps, and enhance international collaboration to better assess chemical hazards;
4. Refine and share chemical risk assessment and risk management globally to promote safe and sustainable use of chemicals throughout their life cycle;
5. Use life cycle approaches to advance widespread implementation of sustainable supply chain management, full material disclosure, transparency and sustainable product design;

6. Strengthen corporate governance: improve chemical and waste management in corporate sustainability policies, sustainable business models and reporting;
7. Integrate green and sustainable chemistry into education, research and innovation policies;
8. Foster transparency: empower workers, consumers and citizens to protect themselves and the environment;
9. Bring knowledge to decision-makers: strengthen the use of science in monitoring progress, priority setting and policy making throughout the life cycle of chemicals and waste;
10. Enhance global commitment: establish an ambitious and comprehensive global framework for chemicals and waste beyond 2020, scale up collaborative action, and track progress.[16]

Superficially, the GCO II proposals seem sensible and likely to deliver very gradual improvement over the long term. However, with unassailable scientific evidence that more than 25,000 human lives are being lost *daily* to chemical poisoning, and in the face of a mountain of fresh evidence of chemical harm that has accumulated since its 2013 report, the 2019 report betrays a chilling lack of urgency. Its language is softer and less candid, its proposals more soothing to industry than its predecessor. Indeed, in defiance of human history it asserts, 'We cannot live without chemicals'. It is not hard to infer that UNEP has been 'got at'.

The GCO II report has alarming gaps. Nowhere, for example, does it state the principle that *all* new chemicals ought to be tested for toxicity and hormonal impact – and that all existing chemicals, so far untested, should also. Indeed, advocacy of universal 'safety testing' is disturbingly silent in the report.

It does note that 'Many chemicals, products and wastes have hazardous properties and continue to cause human health and the environment because they are *not properly managed*' (my italics). Disingenuously, GCO II now attributes the poisoning of multitudes to 'management' rather than to the poisons themselves, and their deliberate, uncaring use. It thus deflects awareness of the global chemical avalanche by attaching the blame to a few bad apples, ill-managed companies, when the entire petrochemicals industry is engaged in expanding the flood. Furthermore, GCO II conveniently ignores the megatonnes of substances emitted by sectors such as mining, farming, construction and manufacturing. And, in its numerous references to chemical emissions, it pays no regard to the need to quantify and inventory the total chemical output of humanity (as is accepted practice for climate emissions, for example) in order to properly assess the risk.

While UNEP may be credited with good intentions, it is clear there is no coherent plan to detox the Earth.

Clean-Up Methods

Despite the unhelpful attitude of the chemical industry as a whole, many individual chemists and some companies – those with a conscience or high ethical and moral principles – have begun work on approaches that show promise for reducing the overall chemical burden on humanity, especially if supported by strong regulation as well as clear market signals from consumers and citizens. These approaches include:

Life cycle assessment (LCA): examines all stages of a product's life from cradle to grave (or cradle-to-cradle as it is sometimes known) – from the original extraction and processing of raw materials through manufacture, distribution, use, repair and maintenance, and disposal or recycling – with a view to minimising pollution and waste at every stage. Besides the chemicals that comprise the object, this assessment includes the energy, carbon and water used in making or distributing it. The advantage to industry is that LCA can also save money. However, it is expensive to set up.[17]

Material flow analysis (MFA): used to study the route of material flowing into recycling or disposal sites, and stocks of materials, in space and time. It links the sources, pathways, intermediate and final destinations of the material. It is used to reduce waste, energy use, unwanted byproducts, costs and chemical contamination all along the chain.[18]

Multi-criteria analysis (MCA): a computational approach that helps to solve complex problems in areas such as production and waste disposal, and to minimise volumes of waste and pollution.[19]

Extended producer responsibility (EPR) or **product stewardship** (PS): an approach that requires the producer to take responsibility for the entire life cycle of their product, including potentially taking it back after its useful life is over and then recycling it safely. EPR is already commonly used for products such as glass bottles, aluminium drink cans and printer cartridges, and is starting to catch on for complex items such as laptops and mobile phones and even industrial chemicals.[20] It is endorsed by the OECD. The virtues of EPR are that it extends the life of scarce resources including metals, minerals and energy, prevents waste at source, and reduces the amount of pollution from subsequent generations of the product.[21]

Green chemistry: a philosophy of chemical production that encourages the design of products and processes that eliminate the use and generation of substances harmful to humans, nature and the environment. For example, it involves creating 'soft chemicals' such as plastics, fuels, pesticides, oils and solvents by biological methods from farm crops or algae instead of from petroleum or coal. A list of these 'safer' chemical ingredients is published by the US EPA.[22] Driven by the quest to cut their own waste, reduce their use of energy and water and avoid mounting criticism by society, many of the world's big chemical firms are showing interest in this approach. The world market for 'green chemicals' is growing steadily, but in 2020 was still only a microscopic $86 million out of global chemical sales totalling $6 trillion – *or 0.0014 per cent* – giving some idea of the industry's sluggardliness in the adoption of benign chemical solutions.[23]

Green manufacturing: means designing and producing products – usually by life cycle analysis or a similar method – that generate far less waste at every stage of their life and which then can be successfully recycled without leaving a toxic footprint. It embodies the idea of 'resource productivity': making more with less.[24]

Green building: means creating and operating buildings and structures that use far fewer raw materials and energy, via the use of life cycle analysis and planned recycling of used or waste materials. Especially it involves the use of materials that are non-toxic, ethical and sustainable and seeking to eliminate chemical emissions from raw material extraction, transport, processing and building operation, demolition and the re-use of construction materials.[25]

Integrated pest management (IPM): a popular concept widely adopted in agriculture to reduce the use of pesticides on the farm and in the food chain by managing crops and food commodities in ways that make them less prone to diseases, weeds, fungal contamination or pest attack. It includes the use of enhanced crop varieties (including some crops produced by genetic engineering) and mixtures, rotations, and approaches to pest control such as the introduction of beneficial insects, pest traps and companion planting or trap crops to divert them.[26]

Regenerative farming: seeks to repair the total agricultural ecosystem – soil, water, biodiversity – and lock up more carbon and moisture in soils, while minimising the use of toxic chemicals and synthetic fertilisers. Potentially it can reduce the 75 billion tonnes of lost topsoil every year. Organic farming and permaculture can also help to eliminate synthetic chemical use.[27]

Chemical leasing: a business model developed in the paint industry where a company does not sell, but rather leases the chemical to a subcontractor to carry out a specific service. This promotes a culture of reduced chemical use to get the job done, rather than industry being driven by attempts to increase the amount of chemicals sold and used.[28]

Industrial ecology: put simply, the idea is to co-locate various waste-producing industries so that what one throws away may readily be used by another industry as a feedstock. The aim is to reduce all forms of waste and find safe, productive uses for it. Again, it involves the study of material and energy flows, and also seeking the best ways to make use of these flows by co-locating productive activities that can benefit from one another.[29]

Zero waste: a set of principles embracing the use and re-use of all products. Zero waste mimics cycles in nature, which wastes nothing. It involves designing and managing products and processes to systematically avoid and eliminate the volume and toxicity of waste and materials, to conserve and recover all resources, rather than burning or burying them. Implementing zero waste seeks to eliminate all discharges to land, water or air that are a threat to Planetary, human, animal or plant health. An attractive feature of zero waste is that it can be – and is being – adopted by individuals to eliminate all wastes in their own home. It applies at every level, from home to suburb to city to industry to nation and thus engages the whole of society.[30]

Risk assessment and remediation[31]: as shown by the Sydney Olympics case, current and historically contaminated sites can be quite effectively cleansed using a range of advanced techniques. The main obstacle to their widespread use is cost – and the reluctance or inability of some industries and governments to pay for the clean-up – especially in newly industrialising countries such as India and China. Technical difficulties (and higher costs) arise in cases where there is a complex mixture of many toxic compounds in soil or water. Nevertheless, the US EPA's Superfund scheme, and its 400 completed clean-ups, has proved it is technically feasible to assess the threats to human health from any of the world's estimated 10 million contaminated sites – and to clean them up, wherever the risk is judged too great. Risk assessment puts the focus of the clean-up effort on the most hazardous sites and substances and reduces costs by avoiding remedial work in cases where contamination is not likely to lead to harm. Typical methods for cleaning up toxic sites include:

- 'dig and dump': the dredging or excavation of toxic material and its removal to a safe landfill;
- 'pump and treat': the process of pumping contaminated groundwater and its cleansing by various technical methods;
- stabilisation: adding special chemicals to contaminated material to stabilise or break it down;
- solidification: use of special chemicals to solidify contaminated material to prevent it from dissolving or moving elsewhere;
- *in situ* oxidation: injection of oxidants to break down susceptible contaminants in groundwater and soil;
- bioremediation: use of special microbes to break down or take up contaminants;
- phytoremediation: use of special plants to absorb toxic metals and other substances from soil or water for removal and safe disposal or re-use;
- electrokinetic remediation: involves the removal of heavy metals from contaminated soils by attracting them to the poles of an electrical current passed through the soil;
- heat treatment to break down toxics;
- ultraviolet light treatment for polluted water and liquids;
- permeable reactive barriers: an underground barrier built across a groundwater flow that absorbs or breaks down contaminants in the water – chemically, biologically or both.

These many and varied approaches to clean-up, ranging from individual sites to whole industries, are proof positive that exposure to toxic substances can be curbed, prevented or, in the worst case, cleaned up after it has occurred; that people do not need to die or suffer unnecessarily so that industry can prosper. That technology has ways to clean up and make safe our living environment – albeit after the event. The same should also apply to before.

The tragedy, highlighted by the extremely low rate of adoption by industry of 'green chemistry' and the very limited adoption so far by cities of zero waste, by farmers of regenerative agriculture, by governments and industry of industrial ecology, is that it is all happening at a snail's pace.

For the time being the decision has been made that it is more profitable that millions suffer and die and that life on Earth proceed towards rapid extinction. Who has the power to change this state of affairs?

Professional Standards

There are reasons why society is reluctant to acknowledge the scale of death and sickness linked to global chemical poisoning.

One of them is that the main professions – chemists, doctors, engineers, farmers, manufacturers – are insufficiently trained and equipped to anticipate and prevent chemical poisoning in the first place. Also, many of these professionals work for, or depend on funding from, industries that have a financial interest in continuing rather than preventing pollution.

Doctors, for example, who are seeing more and more influences of chemotoxicity in the slew of lifestyle diseases they must diagnose and deal with every day, are themselves reliant on the petrochemical industry to supply the raw materials for the drugs they prescribe to try to cure or ease these selfsame lifestyle diseases. Modern medicine is, to a degree, compromised by its dependence for healing on the very process that caused the disease.

Another issue, raised by Dr James Siow of Australia's Institute of Integrative Medicine, is the reduced training received by today's medical students in toxicology during their medical courses.[32] This, he suggests, makes them less well equipped than their predecessors to recognise and correctly diagnose conditions attributable to chronic chemical exposure. In turn, this leads to a tendency to attribute the disease to the patient's own genetics or mental state. And that then leads to further chemical prescription.

Behind this issue is the well-documented concern that many Western doctors are being actively trained, encouraged or induced by pharmaceutical companies to prescribe more chemicals for an already over-chemicalised race of humans.[33] In one spectacular case, drug company Novartis was obliged to pay out $700 million following a lawsuit over bribing doctors with lavish gifts and trips to prescribe their drugs.[34] However, the Big Pharmas are not only selling 'cures', they are also selling sickness. Ray Moynihan and David Henry described this issue at the world's first conference on 'disease mongering' as '... the selling of sicknesses that widen the boundaries of ill health and grow the markets for those who sell and deliver treatments. It is exemplified most explicitly by many pharmaceutical industry-funded disease-awareness campaigns – often designed to sell more drugs than to illuminate or to inform or educate about the prevention of illness or the maintenance of health.'[35]

Large corporations also invest millions in trying to suppress or distort the science around the causes of disease. This practice was first perfected by the tobacco industry – which spent $370 million on research intended to convince the public that lung cancer was caused by the victim's own genes, rather than by smoking. However, the practice has since been widely adopted by petrochemical, pharmaceutical and fossil fuel corporates.[36] Typical tactics include steering the findings of university research through funding grants, controlling the flow of scientific advice to government so it favours commercial interests over public good, attempts to silence and discredit scientific critics, attempts to block publication of adverse findings about chemical products, attempts to divert, muddy and misrepresent necessary scientific discussion of risks, and the outright bribery or employment of corrupt professionals to promote corporate products and views.

Highlighting the trend was a case in which US chemical interests tried to block a hazard study of PFAS,[37] a group of 4000 fluorine-based compounds linked to bladder and liver cancer, endocrine disruption and developmental and reproductive toxicity. These substances are so durable that it is thought that all people now carry them in their bodies.[38] So here is a group of chemicals that are poisoning everyone on Earth, but the public is not to be permitted to know it, nor government to act to prevent it, in order that those who make them may prosper.

Another aspect of the chemicalisation of humanity that is going under the radar is the tremendous mixture of drugs that doctors administer to their patients in order to control the widening array of lifestyle diseases now engulfing society. All drugs have known side-effects, but what remains largely unknown is the effect on human health of the combined cocktail, both of mixtures of pharmaceuticals and of pharmaceuticals interacting with chemicals in the diet, air and water of the patient. All one can say at present is that chemical medicine adds to, and may make worse, the toxicity of the combined human body burden of man-made substances. But we do not yet fully understand the risks of such chemical combinations, and there are powerful forces trying to prevent our doing so.

All these medical shortcomings – the lack of toxicological knowledge, the lack of adequate independent research, the tendency to bow to chemical or pharmaceutical companies and their attempts to bias the science in order to sell more product – render the task of preventing disease in the human population one of immense difficulty. The current paradigm is chiefly to allow diseases (such as cancer) to flourish, then seek chemical 'cures' which usually do no more than fend off a fatal outcome while often prolonging the suffering.

If doctors the world over still take seriously their Hippocratic Oath 'Primum non nocere' ('First, do no harm'), then it is time for major reform of the way Western medicine is both taught and practised around the world, aiming towards the clear goals of firstly reducing the human body burden of chemicals – regardless of whether the exposure was accidental or prescribed – and secondly preventing rather than merely treating disease. For example, since prescribed drugs are implicated in about half a million fatalities a year, on a par with stroke as a cause of death, a good start might be to reduce the death toll caused by chemical medicine itself.[39]

A second profession undeniably in need of reform is chemistry. While many chemists are ethical, well-intentioned people, it is also a fact that many thousands of chemists continue to labour to produce harmful and untested substances, with little regard to their impacts on the lives of millions of ordinary people – even their own children. Also, many chemists appear breezily inclined to under- rather than overestimate the hazards of chemistry, and are often heard to express contempt for the concerns of citizens and families who do not wish to be poisoned, 'because they know better'. There is a distressing ethical oversight in the teaching of universities who, on the one hand, instruct young doctors to 'first do no harm' but, on the other, teach young chemists no such principle.

The Royal Society of Chemistry's ambivalent Code of Ethics makes it plain that many chemists are conflicted over whether they owe their primary loyalty to their employer, be it a corporation or government, or to humanity at large. From now on, it is essential that aspiring chemists both learn and clearly comprehend their ethical responsibilities at the same time as they imbibe their professional knowledge. The time is overdue, therefore, for chemists, too, to take the doctor's graduation oath '*Primum non nocere*'. It is crucial that the chemistry profession as a whole adopt a unified stance that is committed not to the augmentation, but rather to the reduction of the chemical burden in all people and in the Earth system, as quickly as possible. Collectively, the profession must deploy all its intellect, effort and integrity to the tasks of developing low-cost, workable, green chemistry alternatives, and in tackling problems such as chemical mixtures and contamination of the Earth system as a whole, and finding practical ways to mitigate them.

Gender Toxicity

Readers may have remarked, perhaps with irritation, the use of the expression 'man-made chemicals' throughout this book. The words are advisedly chosen. The vast bulk of the world's toxic chemical emissions are generated by men, not by women.

Studies of the gender balance in the chemistry profession globally reveal it to be heavily dominated by males. In the USA, the home of industrial chemistry, for example, the ratio of men to women in university chemistry faculties is four to one.[40] In other countries, the ratio is even more adverse to women. Only five out of 183 Nobel Laureates in Chemistry created since the award was first inaugurated were female. Nine scientists credited with helping to discover substances listed among the periodic table's 118 elements were female. Women are even less well represented among the boards and management of America's forty leading chemical companies, contributing 18 per cent of board members and 13 per cent of CEOs in 2017.[41] While women make up more of the junior ranks of chemists (e.g. 25–51 per cent of undergraduates), they are far from being the dominant voice in the big decisions about what is produced and what is not. The British film *A Chemical Imbalance*[42] explores the gender bias that appears to exist universally in this crucial profession.

Women, on the other hand, are very prominent as world, national and local leaders among the various organisations, consumer bodies and parents' groups calling for a reduction in the toxic burden of society. From this, it is not a long step to a view that the chemical poisoning of humanity is as much a gender issue as violence, sexual persecution, discrimination or other forms of economic, social and political disadvantage imposed by males on females. Chemistry is a masculine profession, and its values are those of traditional males, not of females or even of humans in general. Since the industrial ejaculation of high explosives and poison gas in World War 1, it has sought to satisfy male needs primarily. Today's chemicals, and their characteristics, are the product of male groupthink.

From the parallel fact that men start wars, whereas women almost never do, it is evident that males place a lower value on human life and wellbeing than do women or are, at least, far more tolerant of the needless expenditure of life to gain their ends. This ruthless masculine trait infects the chemistry profession and industry, root and branch, as it does the military, fossil fuels, organised crime, politics, certain sports and some other professions or social groups.

Men are born risk-takers,[43] women less so. Whether it is simply that men are more inclined to underestimate, minimise or dismiss the risks of chemicals – a theory supported by some studies of farm workers – whereas women are more foresighted in wishing to assure a healthier environment for children to grow up in, is not yet subject to proof, but seems like a good doctoral topic for someone. However, this is not a matter of gender politics. It is a matter of human safety and, ultimately, of survival. Once sufficient women are in positions of decision where they can no longer be ushered, coerced or bullied by male collective behaviour, a cleaner, safer, more ethical chemical industry will emerge.

A world in which women lead and share equal power with men will be a far less toxic world.[44]

Can We Adapt?

Humans are nothing if not adaptable, and in recent years scientists have clearly shown that we are continually evolving as our genes respond to new signals in the environment around us. A classic example of this is the physiological ability of adults in cattle-rearing societies to digest lactose, while non-dairying societies are often lactose intolerant: the former groups of people have evolved lactose tolerance to suit their diet and this has occurred in the span of just the last 6000 or 7000 years during which humans have herded cattle.[45] A second example is high-altitude-dwelling people such as the Nepalese and Peruvians, who have evolved blood chemistry that enables them to cope with low oxygen conditions far better than can lowlanders.[46] Finally, there is some evidence that people living in highly arsenic-contaminated conditions may over time evolve tolerance to the poison.[47] Could the same apply to the man-made chemical shower to which we are all now subject? Could we evolve a greater tolerance to toxins in our food, air, water, homes and workplaces?

The short answer, almost certainly, is no. Firstly, the mixtures of substances that come at us constantly, from all sides are far too diverse and too uneven in their dosages for us to make sufficient adaptive genetic or epigenetic changes within a reasonable timeframe, or even over many generations. We might perhaps adapt to a few substances to which we are constantly exposed – but not to thousands of individual substances, or to billions of mixtures. Certain algae, known as extremophiles, have learned to cope with very hot, cold or toxic conditions, but these attributes presumably took millions of years to evolve, and in organisms far simpler than ourselves. In the case of humans, our particular Achilles' heel is the brain and central nervous system, which is especially sensitive to the class of poisons known as neurotoxins – yet is vital to our own personal survival.

Secondly, in order to preserve any beneficial changes in our gene pool, we would have to accept the mass deaths of hundreds of millions people who do not adapt, in order to raise a resistant race through natural selection. In any case, modern healthcare has effectively derailed evolution by keeping alive so many who would not under natural circumstances survive to reproduce. This process effectively preserves not only advantageous genetic changes, but also many disadvantageous ones, in our breeding population and these include chemical sensitivities and susceptibilities.

In short, hoping for timely genetic adaptation of the human species to the man-made chemical overdose is highly unlikely to prove a viable strategy for our ongoing survival on a poisoned Planet. Chemicals will continue to kill us, probably at ever-increasing rates.

Reducing the Burden

Curbing the volume, composition and toxicity of the chemical assault on humanity is the only true solution to the poisoning of humanity and the world. Just as checking climate change involves, eventually, replacing all fossil energy with safer alternatives, the same principle applies to the global chemical suffusion.

After forty years or so, society is slowly coming to accept the necessity for eliminating the carbon-based substances that cause climate change.[48] Public acceptance of the need to end the toxic deluge and return to the healthy environment of our ancestors is emerging far more slowly and hesitantly. Yet the logic is identical.

That millions are dying, that human health is being universally undermined, that our children and the genes we bequeath our descendants are being damaged *at this very moment*, primarily by fossil fuels, constitutes an argument for *immediate* action.

Besides causing the climate to change in ways that threaten the human future, petrochemicals are also responsible for the lion's share of the pollution and poisoning now taking place worldwide. This means that there are now two morally irrefutable reasons – our own health and the Earth's climatic stability – to replace all forms of fossil fuel, plastic and oil-based product use with cleaner, safer, healthier, renewable solutions as quickly as possible. It is, bluntly, a matter of survival.

How we respond to these twin challenges will define the human destiny and the fate of life on Earth.

The largest global source of chemical pollution by volume is released by the mining and mineral processing industry. Here the substances released – spoil, overburden, tailings, drainage etc. – are for the most part less toxic and more easily controlled through good technology, mine design and plant management; the key will be finding the right mix of regulation and market incentives to encourage the global adoption of best practices. The best strategy of all is to accelerate the recycling of all metals, which will dramatically reduce byproduction of contaminants during mining, processing and metal winning. Today's waste streams will become the mines of the future: promising high technologies are already under development around the world to separate and extract valuable metals and minerals from the materials that most people currently regard as 'waste'.[49] On present trends the human population could possibly peak as early as the mid-2060s at around 10 billion[50] (more conservatively, the UN estimates 11 billion by 2100[51]): at this point it becomes possible to meet all human demand for metals and minerals from the waste stream. Demands for new materials and new mines will begin to decline permanently, as they already are for coal and motor fuels. Mineral companies that aim to remain profitable will switch their attention to winning metals from the 'waste stream'. Phytomining and biomining – mineral extraction using plants and microbes – will speed the rebirth of the resources sector as a leader of 'the circular economy', along with sophisticated high-temperature processing of metallic wastes using solar furnaces.[52] So too will the design of manufactured and electronic products that are purpose-built for easy disassembly and recycling of their metals and rare materials; as described earlier, this is already being pioneered by leading electronics brands.

A third massive source of contamination in our lives is hazardous waste. Much of this derives in the first instance from fossil fuels, so the elimination of petroleum, gas and coal use will automatically cleanse our hazardous waste stream substantially. Recycling, biological and advanced thermal treatments and green chemistry, combined with sound regulation and consumer pressure for clean products, can probably eliminate the remainder of hazardous waste creation – along the lines being pursued in Singapore.

A fourth huge source of global pollution is nutrient release, chiefly from agriculture and poorly planned development. The alternative to this is the establishment of large-scale urban food production and the recycling of nutrients, described in *Food or War*.[53] Cities already concentrate two-thirds of the world's nutrients and fresh water, and mostly throw it away as waste. Using this waste, urban farms and novel food systems (such as algae farms, biocultures, entomoculture, aquaponics, food printers etc.) are capable of supplying at least half of the world's food, without the use of industrial poisons or synthetic fertilisers, using *less than a tenth* of the land and water required by today's farming systems.[54] Unlike traditional farming, these systems also have the virtue of being climate-proof. A further food development of critical significance is large-scale ocean (deepwater) aquaculture, using sustainable practices and recycled nutrients to produce the large volumes of fish, algae and seafoods needed to replace the world's failing wild fisheries. This, it is estimated, could potentially supply a further third of the world's food needs by the mid-twenty-first century.

Once urban food production and ocean aquaculture become well established, the harsh economic pressures on traditional farming will be greatly reduced, enabling it to restore its soils, water, agro-ecosystems and the surrounding natural environment. Farmers will no longer be forced to rely on high-intensity chemical-based production systems and can turn to regenerative farming and grazing methods that heal landscapes, end soil erosion, lock up carbon and restore biodiversity. This will in turn help revive the dead zones in our lakes and oceans and bring back lost fishing industries, restore and revegetate the world's rangelands with all their wildlife, and develop more biologically complex and sustainable arable farming systems. Smart farmers everywhere are already embarking on this quest: what they need most at this point is support from smart consumers, smart supermarkets, smart food firms and smart governments with the will to underwrite their efforts through a *Stewards of the Earth* plan.[55]

As the Europeans have already demonstrated, three-quarters of the pesticides used in agriculture and the food chain are needless and can be eliminated without making any difference whatsoever to food supplies or prices – other than improving their health and safety. This constitutes strong evidence that much of the chemical use in world agriculture is currently unnecessary and can be significantly reduced using sustainable systems for managing pest and weed damage – without risking the health of farmers, consumers or even honeybees and birdlife and without incurring significant yield loss in the world's crops and pastures. Furthermore, as the rapidly expanding global farmers' market movement shows, if consumers are willing to pay farmers better to produce clean food, then chemical use in farming can in fact be dramatically reduced, even eliminated, as can the associated degradation of soil, water and loss of life and livelihood in rural communities.

A chemically informed society is the prerequisite to other, similar steps towards detoxing the Planet, including reducing drug dependence (both medical and illegal), replacement of plastics and synthetic fibres with natural or safe substitutes, cleansing the poisoned air of our homes and cities, purifying our streams, lakes and groundwater, producing safe manufactured goods by safe methods and so on.

For all this to happen, however, it is necessary for all citizens to first appreciate the sheer scale of our present chemical burden – and the urgency of curbing it. Only then will we generate the consumer interest, demand and market signals that industries, including food production, need in order to move to new, cleaner and safer methods of production. And only then will governments heed the wishes of the new majority, who will no longer accept that they need to be poisoned in order to eat or breathe. The fact that such a revolution is already under way in renewable energy constitutes powerful proof that it can also happen in renewable food, circular manufacturing and preventative healthcare.

Taking Responsibility

The essential first step in detoxing the Earth is for us to understand and accept our own personal responsibility and role in it.

In one way or another – but mainly through the market signals we send when we purchase food, goods and services – each of us encourages the production of toxic substances throughout the world, whether we know it consciously or not. In many cases we suspect it, but choose to ignore it.

We all want things to be cheap, convenient, safe and useful; while industrial chemicals help to achieve that, we have so far been blind to the risks their overuse entails. Governments and industry, too, prefer 'ignorance', which is why most new chemicals are rushed into global release without proper safety testing and why industry often rejects scientific evidence of harm. As a society we have naively swallowed the view that there is a chemical solution to most of our problems. Such a view promotes the upsides of chemistry while ignoring the downsides. Thus, humanity has gratefully accepted the blessings of chemistry while turning a blind eye to its hazards – hazards that claim one human life every three seconds. The scale of the chemical suffusion and extent of the harm now makes it impossible to ignore. So, by all sharing greater knowledge of the risks, we can together help to change the nature of production.

Building on this first step – the acknowledgement of our shared responsibility for the chemical pollution of ourselves and the Planet – moves us beyond the sterile and unproductive argument over who is to blame, and whether or not industry is at fault for producing what we, the market, order it to produce by our spending patterns. We have to understand that if we demand cheap food, clothes or furniture, then chemicals are an almost inevitable accompaniment, because industry will be driven to produce by the cheapest possible methods, and they are usually chemical.

We need also to grasp that what we save at the supermarket, we spend at the hospital and hospice – or on caring for our chemically damaged children. Considered from such a perspective, is it not wiser to spend just a little more on our food and other consumer goods, to give industry the incentive to produce them by less toxic methods?

The task before us is to provide enough consumers with evidence-based information at a global level to drive change in industry and government by sending the right market signals and economic rewards to industry for doing the right thing. Indeed, the market is probably the *only* way we can discipline or influence industry in ill-regulated regions of the world – that means consumers rejecting its toxic products and preferring clean ones produced by ethical industries that are sensitive to public wishes and truthful about their production systems. Consumer choice is already proving effective as a way to change industries that exploit wildlife, use slave labour, unfair trade and other unethical or unsafe practices. It is a powerful force in the global switch to solar energy, electric vehicles and ethical clothing. Thanks to the internet and social media, advice on how consumers can reshape industry by making sounder choices is spreading round the world at lightspeed.

A cleaner, safer world is a choice we can all make. Together.

9 CLEAN UP THE EARTH

'This is how the thinking layer of the Earth, as we know it today – the Noosphere – came rapidly into being.'

Pierre Teilhard de Chardin, *The Future of Man*

'Colette was our daughter. It is because of Colette – and the countless other children whose lives are being lost to cancer and other childhood diseases that are linked to hazards in the environment – that we are reaching out to you. We want to share with you what parents who have lost children tragically to cancer have learned, and what together we can do about this terrible threat to all children,' say California couple Nancy and James Chuda on their website, *Healthy Child Healthy World*.[1]

The Chudas founded their non-profit public advice website in 1992 after the loss of Colette to a rare form of kidney cancer, Wilms' tumour. Their intention was to memorialise her by helping other parents make better product and lifestyle choices in order to protect their children's health. 'When children are stricken with cancer, you fight for their lives and look for reasons,' the couple explained. 'We asked all along about the cause. People would say, "This is rare. This is non-hereditary." But we had been so careful as parents.' The Chudas could not imagine what the cause of Colette's cancer might be, so they undertook extensive medical tests to see if there was anything in their own background that might have triggered the disease. Other than establishing that it was not genetic, the tests gave them no answers.

'We began to question whether something in the environment had interfered with Colette's gestational development.

We learned ... that it was possible that something Nancy had ingested or was exposed to in the environment during her pregnancy could have triggered the destructive mechanism that caused Colette's cancer to later develop.' It took four years of harrowing inquiry before a scientific study revealed a link between parental pesticide exposure before or during pregnancy and the age of the child at the time of the diagnosis of Wilms' tumour. Among the many troubling facts the Chudas unearthed was the detail that, in the USA at the time, virtually all safety limits for pesticide exposure were based on the amount needed to harm a 155-pound (70-kilo) adult male – not an infant or a baby in the womb.

Then a team of Canadian and Brazilian scientists made a breakthrough. Working in Brazil, where the rates of Wilms' tumour are 'among the highest in the world', and investigating 109 cases of the rare cancer, the researchers concluded, 'Consistently elevated risks were seen for farm work involving frequent use of pesticides by both the father and the mother.'[2] British scientists also found a strong association between the cancer and the children of farmers.[3] Work in the USA later extended this to include auto-workers and welders. All these professions involve high exposure to volatile substances produced from fossil fuels.

'As Colette's parents, we will never forget her bravery,' the Chudas said. 'She taught us not to be afraid to die. She proved to us that unconditional love lasts forever. It is this flame that burns deep in our hearts even today. The morning after Colette died, our close friend and neighbor, Marcy Hamilton, came down the hill to our house. She said, "Colette's favorite color was green. She loved the park. She loved nature. Why don't you start an environmental fund?"'[4]

As time went by, the memory of Colette inspired the creation of the Colette Chuda Environmental Fund (CCEF) to support scientific research into the risks to children from environmental toxics. In 1994 this led to the publication *Handle With Care: Children and Environmental Carcinogens* by the Natural Resources Defense Council, which received worldwide distribution. The following year Senator Barbara Boxer proposed far-reaching changes to the US Toxic Substances Control Act, which included far greater protection for children and pregnant women. Senator Boxer said:

> Nancy and Jim Chuda, despite their grief, chose to turn their own personal tragedy into something positive. They have labored endlessly to bring to the country's attention the environmental dangers that threaten our children. They want to make sure that what happened to their Colette will not happen to another child. No parent should have to go through what the Chudas went through. If future deaths can be prevented, I know we will all be indebted to the tremendous energy and perseverance of Nancy and Jim Chuda.[5]

That perseverance was to bear fruit still more remarkable. Armed with the latest science about childhood cancers and other conditions, the Chudas set to work to make other parents, caregivers, scientists, environmental lobbyists and the media aware of the inherent risks of this chemically saturated world into which children are now born – and what can be done to protect them. This led to the launch of *Healthy Child Healthy World*, an informative, sympathetic and practical website that has reached out to concerned parents and citizens the world over for nearly three decades with trustworthy information, advice and support.

Like the small pebbles that initiate the landslide, families such as the Chudas, the Leakes and the Dengates herald not only a change in how society responds to chemical risks, but potentially a broader and deeper evolution in how humans, as a species, now respond to great and complex challenges.

Global Democracy

Experiences such as those of Nancy and James Chuda, Lisa Leake or Sue and Howard Dengate and other ordinary, loving parents are lending fresh impetus to the cause of raising worldwide consumer awareness about the need to reduce the toxic burden in our lives. This new level of consumer awareness is closely coupled with the rise in 'ethical consumerism', in which informed consumers purposefully avoid products produced in ways that conflict with their moral values – and actively seek out those that meet their standards.

In recent times consumers have banded together round the world to correct major corporations of whose processes, products and behaviour they disapproved. A famous case was the international boycott of the Swiss Nestlé company over its actions in marketing breast milk substitutes in the developing countries, which critics claimed were undermining infant health and survival. This began in 1977 and led to International Code of Marketing of Breastmilk Substitutes being adopted by the World Health Assembly in 1981. The company, and others including Heinz and Abbott, remain under the scrutiny of consumer bodies such as the International Baby Food Action Network (IBFAN) and other organisations, over a range of issues including the sale of drinking water.[6]

A worldwide consumer campaign was launched in 1996 against sports shoemaker Nike over the issue of maltreatment of workers in its Asian factories, dubbed 'sweatshops' in the media.[7] A second boycott was mounted in America in 2018 over issues related to its promotional messaging.

On 24 April 2013, a factory building in Bangladesh collapsed, killing 1134 workers trapped inside. Just five months earlier, a fire at another fashion factory had killed 112 people. In response, consumer boycotts were mounted worldwide, notably against high fashion brands that exploited the ultra-cheap labour and lethal conditions faced by Bangladeshi garment workers.[8] The action prompted debate as to who was most hurt by the withdrawal of consumer support – the companies or the workers.

Ethical Consumer, a worldwide alliance of consumer groups, has been operating at the forefront of global consumer pressure on manufacturers and producers since 1989.[9] Today their website lists thousands of companies and their products that are the subject of current consumer action in fields ranging from food, energy and clothing to health and beauty, travel, banking and technology. They run an 'ethical rating system' covering some 40,000 companies, brands and products to assist consumers in making up their own minds how they spend their money. Products and firms are scored on the basis of consumer feedback. The poorest performers are named and shamed – while the best performers are given a boost.

Rating the efficacy of such consumer actions globally is difficult and controversial, although some scientific attempts are being made. A list of claims for successful consumer actions from 2000 to 2020 appears on the Ethical Consumer site.[10] After more than three decades, the Free University of Berlin's Valentin Beck argues, 'organised consumer boycotts should be regarded as a legitimate and purposeful instrument for structural change'.[11] Indeed, there is mounting evidence that they have helped accelerate the trend to the public adoption of corporate social responsibility (CSR) principles by businesses both large and small,[12] and that 'boycotted firms do significantly increase their prosocial claims activity after a boycott is announced'.[13]

International consumer actions – not all of them are boycotts, by any means – are growing and spreading through the networking power of social media on the internet, as consumers discover they are no longer impotent when faced with monolithic global corporations. Dubbed 'political consumerism' the trend is, in fact, a novel form of international democracy in which people vote for the world they want through their buying power, rather than with a ballot. This reflects a universal movement to address the otherwise utter powerlessness of citizens within the global economy. As corporations replace nation states as the focal power in human geopolitics,[14] global consumer actions are likely to become the chief way that citizens worldwide can express their democratic will and seek to discipline corporate self-interest.

The membership of international online groups involved in consumer action for safer, healthier, more sustainable or ethical products or corporate practices now runs into tens, if not hundreds, of millions of people. Examples include:

- Avaaz is a group established in 2007 with the specific object of 'bringing people-powered politics to decision-making worldwide'. The group's name means 'voice' in several languages, and it was founded to 'organize citizens of all nations to close the gap between the world we have and the world most people everywhere want'. In 2021 it claimed a membership of around 66 million people engaged in almost 3000 separate campaigns. Many of these pertained to chemical pollution, such as campaigns to save honeybees, prevent climate change or end the use of certain pesticides. 'Avaaz has a single, global team with a mandate to work on any issue of public concern – allowing campaigns of extraordinary nimbleness, flexibility, focus, and scale. Avaaz's online community can act like a megaphone to call attention to new issues; a lightning rod to channel broad public concern into a specific, targeted campaign; a fire truck to rush an effective response to a sudden, urgent emergency; and a stem cell that grows into whatever form of advocacy or work is best suited to meet an urgent need,' the group asserts.[15]
- SumOfUs is 'a movement of consumers, workers and shareholders speaking with one voice to counterbalance the growing power of large corporations'. Claiming 17 million members in 2021, it campaigns against 'what happens when powerful corporations get their way'. Specific campaigns include opposition to the use of glyphosate and neonicatinoid pesticides, and to greenhouse emissions.[16]
- 350.org is a global anti-climate change campaign founded by Bill McKibben and colleagues and a prime instigator of worldwide actions against fossil fuel corporations specifically, including its notably successful campaign to persuade investors, large and small, to shun them. The group was also a key mover in the 2018 event that saw 7.6 million people protest worldwide for greater climate action.[17]
- The Safer Chemicals, Healthy Families coalition is a USA-based group consisting of hundreds of medical, social, labour

and civil organisations, businesses and individuals united by 'a common concern about toxic chemicals in our homes, workplaces, schools and products we use every day.'[18] It has three aims: improving policy to better protect the public from toxic chemicals; transforming the marketplace through corporate policies to substitute safer chemicals; and educating citizens and consumers about chemical risk and safety. Its following is thought to number around 15 million in the USA alone.

- With half a century of high-profile campaigning for the environment under its belt, Greenpeace is one of the world's most widely recognised and influential environmental groups. Based in the Netherlands, it has offices around the world, claiming 2.8 million supporters in forty-one countries.[19] 'Greenpeace is campaigning for a toxic-free future where hazardous chemicals are no longer produced, used and dumped into our environment. This includes chemicals which are persistent, toxic, bioaccumulative, carcinogenic and disruptive to human hormones,' the group says.[20]

- Friends of the Earth International (FoEI) is a global federation of seventy-five national environmental organisations and 5000 grassroots groups, founded in 1969 and claiming more than 2 million members worldwide. It runs many campaigns related to chemicals, including publishing maps and databases focusing on pesticides, drinking water contaminants, food imports and PFAS chemicals.[21]

- Founded in 1982, the Pesticide Action Network (PAN) has nodes in five continents and networks some 600 consumer and farmer organisations across ninety countries who are concerned about pesticides. It also has a useful database of 6500 pesticides and their effects.[22]

- Circle of Blue is an independent non-profit journalism service reporting on water issues worldwide, including chemical pollution.[23]

- The WWF (World Wide Fund for Nature) was formed in 1961 and is the world's leading scientific conservation body

working in over 100 countries and claiming 5 million
members worldwide. It has a particular focus on chemical
pollution, its causes, sources and impacts on wildlife and the
ecosystems that support life.[24]

There are now literally thousands of international bodies, alliances
and local groups concerned with the issue of global chemical
contamination, communicating online, in real time, around the
Planet. Listed above are some of the better-known examples – and
they collectively embrace more than 100 million people globally.
To them must be added the millions now engaged in the climate
debate, the food security debate, the nuclear weapons debate, the
extinction debate, the debates over population, pandemic disease,
artificial intelligence, global surveillance and resource scarcity – all
the main catastrophic risks that now confront humanity.

Together these global convergences mark an exciting new
phase in human evolution – a universal attempt to understand,
discuss and solve the common threats we have created to our
own future and to our survival as a species.

This movement heralds the emergence of an entirely new form
of mass democracy – a democracy of the world citizenry in their
fight for survival, health, intelligence and wellbeing against self-
seeking forces, corporate or political, that would deny it.

On matters such as chemicals and climate, humanity is learning to think as a species, for our collective wellbeing and, indeed,
our very future.

Thinking as a Species

At a magical moment in the second trimester of a baby's gestation,
something marvellous happens. The neurons, axons and glia in the
embryonic brain begin to interconnect – and cognition is born. An
inanimate mass of cells becomes a sentient being, capable of
thought, imagination, memory, rationality, feelings and dreams.

Today, the minds of individual humans are connecting, at
lightspeed, around the Planet – just like the individual cells in
the foetal brain. We are in the process of forming, if you like, a

universal, Earth-sized mind. What the French Jesuit philosopher Pierre Teilhard de Chardin termed the 'noosphere'[25] – or sphere of human thought – is becoming incarnate through global electronic connectivity.

A higher understanding – and potentially a higher intellect – is in genesis: capable of understanding, interpreting and solving our problems at supra-human level by applying millions of minds simultaneously to the challenges, by sharing knowledge freely and by generating faster global consensus on what we need to do and how to do it.[26]

At the very moment in our social evolution when governments and existing institutions are seen to be failing to tackle the overwhelming issues of overpopulation, resource scarcity, universal poisoning, environmental loss, nuclear weapons and climate change, a new form of human consciousness is arising that – just possibly – might save us from ourselves.

Even as we now look down on nineteenth-century industry for its filth, cruelty, slavery and exploitation of child labour, future generations will come to view our own era as equally filthy and exploitative of children, because we were willing to sacrifice their health to our own consumer demands and their chemical consequences. Above all we are prepared to sacrifice their intelligence – the very thing that makes us human.

To avoid such a judgement of history, citizens the world over need to comprehend the almighty scale of the poisonous flood engulfing us all, a flood that kills 25,000 people every day – and the plain fact that it is within our power to abate it, if we so wish. It is within our reach to bring to a close what our descendants will undoubtedly view as a Toxic Dark Age of preventable sickness, suffering and ignorance.

The means for achieving this new Enlightenment already exist: the internet and social media. As of mid-2020 the internet reached 4.6 billion citizens of Planet Earth (59 per cent of the population)[27] and was on track to include everyone before 2030. Through the internet and social media, for the first time in our history, humans as a species are sharing thoughts, ideas, problems and solutions around the Planet instantaneously, across almost all the political, religious, cultural, racial, national, economic, social and physical barriers that have divided us and hindered understanding in the past.

Humans are in the process of becoming cognate at the species level – and this is an awe-inspiring and hopeful moment in our evolution.

For the first time, this novel medium of communication has opened the way for a new human connectedness via shared thought, knowledge, values, consensus and joint action on issues that imperil our wellbeing and our future. Moreover, it is happening in real time and its ideas are transmitted at lightspeed, or close to it. For all our individual differences, the road is now open for us to work together as a species.

True, the internet is rife with trivia, rubbish, ignorance, abuse, hatred and 'fake news' – but it is also laden with science, common sense, idealism, hope, trusted fact, generosity, love and well-intentioned activity. Like every human mind, it carries thoughts that are generous, wise and altruistic – and thoughts that are mean, ignorant and petty. As individuals, most of us try to steer our lives in the direction of good thought – and there is no sound reason why, globally, Planet-wide human thought should not also obey this pattern. Such values are, after all, already embedded in most of our collective institutions – religious, ethical, social, educational and governmental.

Individuals who value a healthy future for themselves and their children will network online with like-minded people, who also care about their children and about the future of humanity. Through the power of social media, especially, such people will soon influence many who fear it is beyond their ability as individuals to change our world for the better. In truth, it has never been more possible.

If all the groups of consumers, citizens, parents, victims, farmers, scientists, environmentalists – and others who are concerned about the harm we do to the Earth we inhabit – were to join hands in cyberspace to share information, ideas, advice and mutual support for a cleaner, healthier and safer future, this would be a more influential gathering of people than any political, religious or national movement now or in the whole of human history. If they were not only to pool ideas, information and thoughts but also to inform purchasing decisions among several billion consumers, this would send strong signals to industry, by rewarding clean, safe products and industrial processes, and penalising polluters and poisoners with loss of profit.

Everyone on Earth is a consumer. We all carry a chemical burden. Many people are, or become, parents. Most of us are potential cancer victims. And all of us are having our minds, bodies and genes subtly distorted by chronic chemical exposure, every single day.

By uniting, sharing knowledge, educating one another, and choosing clean industries and businesses, we can make a very large difference – one so large, in fact, that most political, religious, governmental and commercial entities will want to be a part of it. Clearly such a global group will only thrive if its objectives are above ideology, prejudice and vested interest, and especially nationalistic and corporate self-interest. If a clear and present danger to the health of all humans and their offspring into the future cannot unite us to take collective action about this menace to our existence, then perhaps we are not the *Homo sapiens sapiens* we fancy ourselves to be; perhaps we should be looking for a name more suited to an unwise, self-destructive species.[28] There are plenty of global bodies – such as the International Red Cross/Crescent, Oxfam, the UN and World Health Organization, Avaaz and the IPCC – that prove it is possible for humans from diverse backgrounds and beliefs to come together on matters of mutual concern and responsibility.

To detoxify the Earth is simply an undertaking on a larger scale, involving all of us and making full use of the most pervasive means of knowledge sharing ever devised.

A Global Detox Alliance

A core finding of this book is that we must build a Global Detox Alliance to:

- Share information universally about toxic chemicals of concern and their health impacts as revealed by independent, peer-reviewed scientific evidence.

- Share information on which products are dangerous and should be avoided.
- Share information about which products and processes are clean, healthy and safe and should be encouraged and rewarded.
- Educate parents, consumers and children on how to choose the safest products from the array on offer.
- Accelerate the global sharing of scientific knowledge about contaminants among government regulators, international treaty bodies and citizens around the world by reporting new issues and substances of concern.
- Press for increased public funding of research into chemotoxicity and for the mandatory safety testing of *all new and existing* chemical substances and mixtures.
- Demand the creation of an open, transparent global chemical inventory.
- Lobby for stronger support for preventative healthcare, to supplant the costly and pervasive chemical therapy model.
- Promote and reward the uptake of 'zero waste', circular economics and similar philosophies by local communities, industries and governments.
- Establish a 'Forbes 500'-like list of the world's cleanest companies, to convey the message that health and safety are as important as profit, and to acclaim and honour the best performers as role models and market leaders for others to follow and emulate,
- Work with industry to develop clean codes of practice and ethical standards that meet consumer and civic, not just industry, needs and independent scientific standards.
- Provide evidence-based educational material to schools worldwide about the risks to children's health from chemicals (including illegal drugs, alcohol, tobacco, certain processed foods etc.) and thereby help children to educate their parents.
- Create smartphone 'apps' (applications) and a public ratings system that advises consumers which products are safest – and which to avoid – as they shop.

Such an alliance would not engage in consumer bans or boycotts, physical confrontation, lawsuits or other direct action against industry or science; to do so will only entrench mutual mistrust and opposition, delay the move to clean production and drive industry into greater secrecy and into unregulated parts of the world. Clean-up will do best if founded on principles of cooperation, consensus, openness and equality between society, industry and government.

Such a group need not even be a formal organisation – but rather a networked movement of like-minded, well-intentioned bodies and individuals, with a guiding set of principles and a common forum for discussing them. In fact, if operated on respect, information sharing, common cause and rational persuasion, such an alliance will be far more influential and effective than a formal lobby group or NGO. It will also be a lot harder for its critics to discredit it, defuse, divert or shut it down (in the same way industry and political lobbies have already sought to silence and mislead public debate over issues such as global warming, acid rain, tobacco, environmentalism and DDT by gagging public and scientific institutions). A global alliance can equally apply pressure to local and national governments to decide where they stand in the choice between public health and corporate self-interest.

It need not embrace every consumer on Earth. A simple majority will do. Or even a groundswell significant enough in number to hold the attention of industries, governments, media and stockmarkets everywhere.

Ideally, a Global Detox Alliance ought not be an official body or anything that smacks of 'world government'. If the people of the world do not wish to be poisoned, they can express their wish through knowledge sharing and their own behaviour in the market without need of authority structures. A people's movement is more the mark.

A united effort to clean up the Earth can become the first true manifestation of global grassroots democracy, founded upon shared values and common interests – of a kind no existing governmental structure anywhere in the world can yet deliver.

If as a society we truly value our health and that of our children for generations to come, then the issue of toxic exposure is a fundamental concern, wherever we live or whoever we are. As a concept, the risk of being poisoned is not difficult for the individual to grasp. It therefore has a better shot at building a Planet-wide consensus than other catastrophic threats (including climate) – enabling us to take a united step into a new human future of self-governance by mutual agreement.

Cynics, pessimists and vested interests will certainly dismiss this as naive idealism, but such criticisms are always levelled by lazy minds at new and challenging ways of thinking about great issues. It will also be the target of those who cling to the atrophying, historical power structures (such as the armed nation state, fossil fuels and buccaneer corporates) which are so plainly failing to deal with the ten rising threats to human health, welfare and existence.

A Matter of Science

Over several decades and thousands of individual pieces of research, countless scientists have managed to demonstrate, beyond a shadow of a doubt, that chemical contamination has burgeoned from a local issue to a global one. However, while international regulatory agreements have been formed and strategies enunciated, as yet no world scientific body has been tasked with quantifying or understanding Earth system contamination on anything like the scale and detail devoted to climate change, pandemic disease or other catastrophic risks. On this, science has – so far – let humanity down.

Lacking this essential information, the issue of Earth system poisoning and the Anthropogenic Chemical Circulation has been underestimated and ignored by governments, by industry and by society at large. It has been treated as a mass of second-order issues, involving individual chemicals, companies, local events or diseases rather than the universal menace it now presents.

Contamination by the chemical products and byproducts of human activity is one of the most pervasive and far-reaching of our impacts on the Earth and on our own health and wellbeing. The fingerprints of anthropogenic contamination are now to be found from the stratosphere to the deep oceans, from pole to pole, in many forms of wildlife, in all modern societies, the food chain and in most individuals, including newborns. Chemical contamination is of equivalent significance with climate change, the Sixth Extinction, the nuclear arms race and all the other major impacts of human population growth and overdevelopment. Like them, it will define our future.

Fortunately, a new venture – *globalCARE* – is stepping up, to remedy the situation in science. *globalCARE* is a worldwide scientific endeavour launched at the international *CleanUp* conference in 2013, with support of scientists from many countries, to define, quantify, set limits to, help clean up and devise new ways to curb the growing chemical assault on human health and the biosphere. 'We envisage this as an international alliance of leading scientific, government, industry and community organisations and individuals dedicated to making ours a cleaner, healthier and safer world,' says its founder, Australian Professor Ravi Naidu. Subsequent world conferences were held in 2017 in India and in 2019 in Korea.[29]

globalCARE states as its purpose:

1. To better understand the nature, extent, circulation and impact of Earth system contamination on human and environmental health and wellbeing.
2. To assemble global data for international, national, industry and health bodies engaged in reducing the impact of contamination.
3. To develop and assist the adoption of cost-efficient technologies to assess, clean up or prevent contamination.
4. To share scientific knowledge and technologies for assessing, cleaning up and preventing contamination, locally and globally.
5. To educate consumers, industry and governments about contamination, its pervasiveness, its adverse effects and how to prevent them.

6. To create value by developing beneficial uses for contaminated land.

7. To prevent disease by fostering cleaner industry and contamination-free food production.[30]

Professor Naidu explains:

> The initiative seeks not only to define the extent of contamination at international scales, but also to develop and share cost-effective, workable solutions which can be readily adopted by industry, governments and the community ...

> These include further developing and disseminating the concept of 'green production' – the production of goods and services without accompanying risk of contamination. globalCARE is a worldwide knowledge network, performing new scientific research, aggregating existing knowledge, developing novel assessment and clean-up technologies, advising governments and industry on ways to improve existing regulation or industry practices, training high-level experts and sharing information about ways to reduce anthropogenic contamination in all facets of human society and the natural environment.

Among globalCARE's planned tasks are developing a 'stocks and flows' model of global contamination, efforts to establish safe global boundaries for key pollutants, identification of urgent issues and areas for action, investigation of the combined effects of contamination on human health and the environment and the international sharing of solutions. The initiative will work in partnership with other major scientific programmes (such as the seven major infant toxicity studies referred to in Chapter 3) and with leading universities and corporations.

'There is a widespread lack of awareness among governments and societies about the current scale, pervasiveness and risk to billions of people from contamination in the Earth system,' Professor Naidu says. 'Indeed, so complex is the threat that we do not yet fully understand it or how best to curb it. There remains a grave lack of international institutions committed to tackling it – and a lack of the science needed to underpin their work. There is also a lack of a real appreciation of the many benefits and economic, social and environmental gains to be obtained from cleaning up our world. The challenge is urgent – and we need to start now, working at global as well as local level.'

The first-ever attempt to build a worldwide chemical inventory, described in Chapter 1,[31] represents an essential first step in understanding and curbing the deluge. As does the development of an effective Global Framework for safe chemical management to replace the expired SAICM.

Absolutely critical is the development of a body equivalent to the Intergovernmental Panel on Climate Change (IPCC) to measure, monitor and oversee a worldwide endeavour to clean the place up – a Global Panel on the Anthropogenic Chemical Circulation (GPACC) – a formal international partnership involving all the world's governments, built around the ideas propounded by *globalCARE*.

A Matter of Ethics

A dimension of scientific failure over the contamination of the Earth system in urgent need of amendment is the ethos of the chemistry profession worldwide.

Chemistry is science's senior discipline and its conduct has powerful sway over science as a whole. While nobody doubts that most chemists work from good motives, with the intention of bettering humanity, their failure as a profession to accept or address the cumulative harms inflicted by the totality of human chemical emissions is an act of denial in urgent need of reform. If the aviation sector, for example, were to say 'air safety is not our concern', it would be subject to wide public condemnation and probably closed down. To tolerate 9 million deaths and 86 million injuries a year attributed to chemical emissions is unacceptable by any reasonable human standard – and must change.

A good start will be to include an ethics component in all university chemistry courses, requiring practitioners to consider the longer-term and wider consequences and impacts of their work, especially when adopted at industrial scale. In future all graduating chemists should be asked to take the doctor's oath,

Primum non nocere – to help, but first, to do no harm. This concept, known in ethics as the principle of non-maleficence, is attributed to the Greek physician Hippocrates,[32] but was proposed for modern medical practice by Thomas Sydenham in the seventeenth century and Thomas Inman in the nineteenth century.[33]

I here propose the adoption of an *Oath To Do No Harm* for all modern practitioners of chemistry and related disciplines.[34]

Chemists should be oath-bound to consider the toxic consequences of their novel products, not just in the test tube, but at global scale and in consequence of mixing.[35] They should be required, by their professional code, to safety test all novel compounds and, indeed, all existing widely used compounds. The profession as a whole should feel an obligation to reduce, with all the skill, intellect and energy at its disposal, the cumulative toxic flood to which humanity is now subject. Chemists cannot be held responsible for the explosion in human numbers or the vast demand for material goods that has unleashed the chemical flood – but they could have taken greater precautions against bad outcomes – and they share a heavy responsibility for cleaning it up. Just as the oil and coal sectors knew more than half a century ago they were destroying the Earth's stable climate and did nothing, the chemicals sector too has been fully aware of its own detrimental impacts on human health and the environment, but has on the whole sought to downplay, dismiss or deny them.

In the end, it comes down to leadership: are the leaders of today's chemicals sector more concerned about profits and funding than they are about human health and wellbeing? Or have they the moral courage to admit that, with the overgrowth in human numbers and material demands, what might have been acceptable in the 1950s now represents a catastrophic risk to humanity, and indeed to life on Earth, which must be contained and controlled as a matter of urgency? Importantly, the chemical *industry* must be persuaded to cease its efforts to distort the science, muddy the flow of information to the public, slander its critics, and oppose or weaken laws and regulations intended to protect the health and wellbeing of society.

This requires moral leadership of an altogether different calibre to that seen today. Good chemists and young chemists must speak up – and lead their profession to a cleaner, safer future. The inclusion of far more women among the profession's leadership will accelerate the change.

A Matter of Rights

Every person in the world has a right to life, liberty, personal security, to marriage and family, to equality, to work, to education, to the law, to elect a government, to freedom of opinion or belief, to asylum. These are just some of the rights available to each of us under the thirty articles of the *Universal Declaration on Human Rights* (UDHR),[36] though it is true that many governments still fall short of these prudent and fair requirements when measured by how they treat their own citizens and others. Nevertheless, such human rights represent a key aspiration for all the world's inhabitants in the twenty-first century, as well as an important yardstick by which governments and regimes may be measured and judged by their citizens.

It is therefore more than a little disturbing to find there is no *Human Right Not To Be Poisoned*.

A child born today may enjoy all of the rights listed under the UDHR – but not the right to a full intelligence, to undamaged genes, to a life free of cancer, mental or reproductive dysfunction or other attributes increasingly linked by science to lifelong chemical exposure. The lack of such a right, in the presence of a right to leisure, to social security or to cultural participation, bespeaks a remarkable blind spot in the contemporary conscience, deriving either from an acute lack of awareness of the scale of the problem or from a general desire to hide from ourselves knowledge that is unpleasant, distasteful or disturbing, especially if it means having to change the way we produce material goods, food, energy and how we go about things in our daily lives.

In Article 5, for example, the *Universal Declaration* proclaims that every human has a right not to be tortured. Although this is a right that, presumably, refers to only a small percentage of the world population at any one time, there is no articulated right for people to be safe from the flood of toxins or suspected toxins that now engulfs the entire human species, cradle to grave. To be safe from an assault that we know kills millions annually and impairs the health of tens of millions more – often in ways that might well be deemed torture if you caught someone deliberately inflicting them.

Likewise, everyone has a right to education – but apparently not a right to the inborn intelligence that would enable us to take full advantage of it.

We all have a right to security of person – except when we inhale, ingest or absorb something toxic from the throbbing industrial machine that now encircles the globe.

We have a right to equality before the law, except when it comes to challenging the producers of toxins, who so often use the law as a shield against their critics and against the necessity to change their dangerous products or processes.

We have the right to marry and have a family – but not one whose health, wellbeing and future are protected from the contaminants, both legal and illegal, now circulating in the Earth system.

We have a right to equal access to the public service of our country, except in cases where the public service, often acting under political pressure, takes the side of a contaminator against the contaminated, of corporations against its citizens, and frustrates their reasonable safety concerns.

We each have a right to a job – but no clear right not to be poisoned while doing it.

We each have a right to 'a standard of living adequate for the health and well-being of himself and of his family', but not a right to a standard of living as free from toxins as past generations of humans have enjoyed.

The mere existence of a human right does not prevent that right from being abused in a great many cases – but it does establish a world standard of behaviour for humanity in general. It constitutes powerful moral suasion on corporations, governments, communities and individuals not to breach it, spurred by the threat of exposure if they do.

One of the principal difficulties in attempting to enshrine a *Right Not To Be Poisoned* is that there is at present not a single government or regime on Earth that could truthfully claim to uphold it. Even the most scrupulous, strict and efficient administrations have no solution to the twin problems of Earth system contamination and toxic mixtures – and, as the UN Environment Programme has pointed out, administrations are still broadly in ignorance of the toxicity of most of the chemicals in regular use – or even now being introduced – in their own jurisdictions. For this reason, any proposal to introduce such a Right may face opposition from nation states, reinforced by industries and interests that profit from the *status quo* and are loath to change.

But that does not mean that concerned citizens around the world should not seek the instatement of such a Right, or seek to arouse others to an awareness of its necessity. Rights, we know from history, are not to be had for the asking. They are to be argued, campaigned and fought for, often against bitter and entrenched opposition, usually over many years and sometimes generations. However, they represent a standard to which all people, nations and corporations may aspire. The existence of a Right will give us all hope for a better future.

If we do not have a *Right Not To Be Poisoned*, there will never again be a day in our history when we are not.

Detox the Earth: A Ten Point Plan

If we wish to curb the toxic burden for ourselves, our children and for all life on Earth, we must take agreed universal action. The following ten points offer the basis for a plan:

1. Form a global alliance of people, institutions and businesses concerned about detoxing the Planet, to spread awareness, motivate the uptake of clean products and production systems and educate citizens to become 'clean consumers'.
2. Campaign for a universal *Human Right Not To Be Poisoned*.

3. Establish a new international scientific body to measure the full extent of human chemical emissions, assess their toxicity and impact, monitor change and oversee the task of cleaning up the Planet.

4. Press for the universal safety testing of all new and suspected chemicals. Share the findings openly and globally. Maintain an open global toxic substances register.

5. Press for the replacement of all coal, oil, gas and other fossil fuels with clean energy and with non-toxic feedstocks for industry.

6. Banish known toxins from the food chain, water supply, air and wider environment through informed consumer choice and regulation. Increase scrutiny of suspected toxins.

7. Press for a priority policy of disease prevention in medicine, over chemical cure. Educate healthcare workers to recognise, diagnose, report and prevent diseases resulting from chronic or acute chemical exposure, and educate the public about the risks and solutions.

8. Train all young chemists, scientists and engineers in their social and ethical responsibility to 'Help. But first, do no harm'.

9. Educate our children to choose wisely among products and services those that are safest and least toxic. Empower them to educate us.

10. Empower and reward industry to make profits ethically, by producing clean products that do no harm. Encourage universal adoption of stronger clean industry codes, recycling, zero waste, green chemistry, clean-up and so on.

There are plenty of other things we could do, but ten are sufficient for the purpose of starting a movement to detox the Earth whose aim is to herald a clear and present danger to ourselves and to future generations – and also to set forth the inspirational possibility of our overcoming it, together, in ways that lead to better health, greater prosperity and to a cleaner, more sustainable and a fairer world.

A toxic Planet and a poisoned people may be averted if we, the Citizens of Earth, so will it.

10 PREVENTING CATASTROPHE

'Human history becomes more and more a race between education and catastrophe.'

H. G. Wells, *The Outline of History*

Though larger than global warming in emissions and more deadly than either famine or pandemic disease, the poisoning of Earth is but one of ten catastrophic threats brought on humanity by our own numbers, actions and heedless overuse of our Planet's resources. It does not stand alone as an issue but is an interconnected part of a growing menace to the very existence of the human species.

To make the distinction clear, a catastrophic risk is one that threatens all or most of civilisation with disaster; an existential risk is one that threatens our actual survival as a species. Of the ten, only two – climate change and nuclear weapons – qualify as specific existential threats to humanity in that, on their own, they can wipe us all out. The other eight, such as pandemic disease, food insecurity, ecological devastation and global poisoning, represent catastrophic risks to civilisation at large. Taken together, however, all ten risks constitute the greatest existential emergency ever to face humans in the million years of our tenure on the Planet. Global poisoning with chemicals is not merely a catastrophic threat to human health and wellbeing; it also gravely undermines our fitness to survive.

The ten risks are described in *Surviving the 21st Century*[1] along with the science behind them, the causes and what humanity as a whole and we as individuals can do about them. These risks are:

- Decline of key natural resources and an emerging global resource crisis, in water, soils, forests and the oceans especially.
- The widespread decline and collapse of natural ecosystems that support all life, including our own, and the sixth mass extinction of wild animals and plants.
- Global heating, sea level rise and increasing turbulence in the Earth's climate affecting all human activity, especially our ability to produce food.
- Universal contamination of the Earth system and all life by emitted chemicals – the topic of this book.
- Rising food insecurity, declining nutritional quality and increased risk of conflict.[2]
- Nuclear weapons and a new global arms race.[3]
- Megacity collapse, linked to population overgrowth and resource failure.
- The increasing frequency of outbreaks of new and old pandemic diseases generated by human activity.
- The introduction of uncontrolled and dangerous new technologies.
- Widespread delusion and ignorance across society about the scale and nature of the risks we now face, leading to inaction.

The central message of *Surviving the 21st Century* was that all these risks are interconnected. They cannot be solved on their own, or one by one. To secure the human future, they must all be solved together, in a systemic way, and by methods that *make none of them worse*. The following commentary explains how the chemical threat interconnects with the nine other risks and points to possible ways forward.

1. **Climate change** is the largest recognised impact of unrestrained human chemical emissions. It is driven mainly by the burning of fossil fuels, but also by a host of other insults including fertilisers, pesticides, plastics, synthetic textiles, refrigerants, HFCs, PFCs, methane, nitrous oxide, nitrogen trifluoride etc.[4] However, warming chemicals account for less than one-quarter of total human chemical

emissions, leading to the question: if greenhouse chemicals can cause so much world havoc, how great is the risk from all the other substances released by human activity? Unquestionably, the elimination of climate emissions (involving the closure of the coal, oil and gas sectors and their replacement with renewable energy and other substitutes) will also eliminate the primary source of toxic chemical pollution on the Planet and a major risk to human life and health. It must therefore proceed with added urgency because of the millions of lives now being sacrificed. However, care must also be taken not to simply exchange one form of pollution for another, such as nuclear radiation, toxic mineral processes or new, untested substances. Accelerated action on climate will help significantly to detox the Earth – and accelerated action on global pollution will help limit climate change.

2. **The Sixth Extinction** of life on Earth has several main drivers, including land clearing (for industrial agriculture), urban expansion, wild harvesting and the poisoning of all wildlife by human chemical emissions, notably pesticides and endocrine-disrupting chemicals. The chemical flood therefore plays a key role in the world ecological crisis and the rendering of Earth less habitable for humans and other animals. The withdrawal of these chemicals from use will not only save millions of human lives but also reduce extinction risk for critical life forms such as insects, which support birds, frogs, fish and other animals. Ecosystems support the wild world in supplying the clean air, water and food which are also indispensable to human survival. It makes no sense to poison them. Reducing the chemical flood will arguably constitute a major step towards ending the Sixth Extinction and restoring the world's life-giving ecosystems to health.

3. **Food security**. Chemistry is now intimately interwoven with the production and processing of more than half of humanity's food supply. Because chemicals are cheap and easy to use, Western agriculture, food processing and

packaging have developed an addiction to them which, additively, poses sharply escalating risks to humans, wildlife and to the environment. This has fuelled a dangerous paradox, in which chemicals are key to maintaining a high output of poor quality industrial food, yet are increasingly implicated in both chronic and acute poisoning, and in the rise of lifestyle diseases – among them diabetes, obesity and cancer. It is now imperative to purge the food chain of chemicals with known toxicity to humans and wildlife and replace them with alternative technologies or softer chemistry. The existing model of world food production is unsustainable (as explained in *Food or War*) and must be replaced by one that consists of (i) regenerative farming, (ii) renewable urban food production and (iii) deep ocean aquaculture. This novel food system will minimise the use of chemicals and will help greatly to detox the Planet. A renewable food supply will also help end the Sixth Extinction and will reduce the threat of war.

4. **Nuclear weapons.** Chemistry is a fundamental component in the design of materials used to make advanced nuclear weapons and is thus a prime contributor to Armageddon. The only solution is to ban nuclear weapons, along with all the materials and processes by which they are made, and to eliminate all supplies of fissile material, as proposed in the UN Treaty on the Prohibition of Nuclear Weapons. Chemistry can also play a vital role in making these substances safe and recycling them beneficially. However, the ethics involved in the creation of weapons of mass destruction must also be the subject of serious self-examination by the chemistry profession: doctors do not plan the mass death of humans on a global scale. Some chemists and physicists do – and it is time they were held to account and required by their profession and society to behave in a more humane and ethical manner.

5. **Resource scarcity.** Growing shortages of soil and clean water, the loss of forests, the decline in global fish stocks and ocean health as well as scarcity of key minerals are all linked

in different ways to the chemical flood. Major drivers include industrial agriculture, industrial forestry and fishing, climate change, the universal pollution of fresh waters and the oceans, the poisoning of wildlife including fish; all these sectors depend on or derive from chemistry. The harmful role of chemicals in each of these looming resource crises needs to be clearly mapped and mitigated. On the other hand, the role of chemistry in recycling, in building the global circular economy, in developing safe and sustainable alternative materials, in ending pollution and locking up poisons and in cleaning up contaminated regions needs to be ramped up. This will involve the chemical profession and industry shifting from a harm-inflicting to a healing, regenerative mindset, helped by universal consumer demand to motivate it. Female leadership may be essential to this transition. Consumer support for green chemistry certainly will be.

6. **Megacity collapse**. Megacities are the fruit of the human population explosion which, aided by modern chemistry in food and medicine, now far exceeds the ability of the Earth to support it.[5] Megacities are themselves living far beyond their means. As described in *Food or War*,[6] no megacity can feed itself, and all rely heavily on long, chemical-driven food chains to supply their needs. Failure in those chains, due to climate change, resource depletion and eco-failure, spells disaster for billions of people. The solution is 'renewable food' – the adoption of advanced urban food production, which uses almost no artificial chemicals and depends on recycling water and nutrients, and deep ocean aquaculture.

7. **Pandemic disease**. Pandemic diseases arise chiefly from human destruction of wilderness for industrial agriculture and urban sprawl, combined with the overgrowth in our population. These bring us into close contact with the animal hosts of new and unknown diseases, allowing them to jump species. They then spread by world travel, urbanisation, food chains and human behaviour. Chemistry is involved in all facets of their genesis – and is also deeply

engaged in trying to combat them with new drugs, vaccines and protective materials. The role of chemistry in helping to start pandemics has not been deeply considered, and must be, for the sake of preventing new ones in future. Its role in developing safe, sustainable solutions for existing pandemic diseases must be accentuated.

8. **Uncontrolled technologies.** These include things like the spread of universal surveillance by governments of their citizens, and corporations of their critics and consumers, the adoption of artificial intelligence, killer robots, nanotechnology and the engineering of new life forms.[7] All of these are proceeding without public oversight or permission and, like chemicals, are now being released onto the world without control and without due consideration of their larger consequences. Chemistry is an essential input to all of them and so bears moral responsibility for helping to limit the threats they pose to the human future. Universal surveillance (enabled by quantum computers, AI etc.), in particular, can be used to silence concerned citizens and gag scientists who wish to warn about the harms inflicted by the institutions that wield these tools of tyranny. The suppression of scientific truth and free speech has been common practice by the chemicals sector since Minamata. Reform and public transparency are now imperative.

9. **Mass delusion.** Perhaps the most dangerous element in the ability of human civilisation and our species to survive in future is our capacity for self-delusion – for believing whatever we fancy, in spite of the evidence. This risk is discussed and explained in *Surviving the 21st Century*. Delusion is the mortal enemy of rationality and truth, and hence of our ability to survive the real threats we face, including the chemical flood. The chemicals sector – along with fossil fuels, tobacco and pharmaceuticals – has been a skilled disseminator of misinformation about chemical risks, and hence a feeder of public beliefs that they are an insignificant or, at worst, a second-order risk to human life. The science flatly contradicts such a view. It is time that

society based its assessment on independent scientific fact, not on self-serving industrial propaganda, denialism or attempts to distort the science. Only if told the truth will consumers and citizens fully understand the necessity to choose the products that empower clean chemistry and eliminate the substances that will poison generations of humans yet unborn.

In summary, our ability to survive and prosper through the twenty-first century and beyond depends on how successfully we can solve the ten interconnected threats, the gravest existential emergency our species has ever faced.

Humanity's ability to inflict mass harm on itself has been accelerating for the past hundred years. Worldwide trends in population, industry, politics, warfare, climate, environmental destruction and high technology have delivered an entirely new level of risk, one unseen in history. The risk is global, complex and potentially existential.

The answers lie not only in sound governance and ethical corporate and scientific behaviour but also in the willing combined actions of billions of individuals in their daily lives. Much of our present behaviour has to change – but changing it will bring fresh opportunities for health, prosperity, work and fulfilment.

Together, we now face unarguable proofs that our combined chemical outpouring threatens human civilisation, placing in jeopardy the health, happiness, intelligence and wellbeing of all.

Now is the time for us to clean up the Earth.

Together.

POSTSCRIPT:
A CAUTIONARY TALE
FROM DEEP TIME

It was a fine Tuesday morning about 2400 million years ago. A faint but increasingly vigorous young sun shed its silvery radiance through gaps in the roiling clouds and layers of mist that still enshrouded the young Earth, bestowing light and heat to warm the sultry waters of a virulently tinged sea. Within those waters, vast mats and clumps of vivid green algae and purple microbes basked in the fierce sunlight, soaking up its energy and quietly using it to make and digest useful food. Every so often one of them would emit a tiny, satisfied belch. Invisibly, tiny molecules of oxygen entered the water and, finding nothing much to cling to, joined their companions to form a bubble which rose slowly but resolutely to break the oily surface of a stagnant world ocean. As far as the eye could see – had there been any eye to see it – the watery world was all bubbles, thrusting through the primordial sludge of early life, bursting into the atmosphere like the expiring fizz from a bottle of lukewarm soda water.

In many ways it was a beautiful Eden: the sky had a greenish tinge, the ocean formed a rainbow swirl of reds, blues, greens and yellows reflecting the teeming life and the patterns of light that played upon it as pulses of sunlight rippled through the racing cloudscapes. The continents were dusty reds, yellows and greys, devoid of all vegetation and stripped to the bone by the sandblast of cyclonic winds and the torrential rains that levelled mountain ranges higher than the Himalaya in the twinkling of a geological eye. Life was somnolent but rich for the countless billions of stromatolites, blue-green algae and other microscopic organisms which burgeoned in muddy pools, tepid lakes and tidal wetlands, cloaked the surface of bays and inlets, slurping the nutritious richness of minerals dissolved from volcanic springs and seabed smokers.

For more developed life forms, such as ourselves, it would have been a deadly Garden of Eden: simply drawing a breath from the atmosphere rich with nitrogen, carbon dioxide and water vapour but almost devoid of precious oxygen, would kill you in no time. Even if you could breathe, you'd have been fried by blistering ultraviolet rays from the young sun sparkling through the interstices between the clouds.

Stromatolites and blue-green algae were the grandchildren of Luca, our 'last universal common ancestor', a creature dwelling so far back in the misty past as to be almost a scientific legend. Luca itself evolved on – or came to – the Earth about 4 billion years ago, as soon as the first torrential rains had filled the ocean basins, formed lakes and rivers and created a suitably damp home. Luca was a lean and hungry beast, largely given to devouring its own kind for the organic carbon it needed to survive.

Then, about 500 million years in, a miracle occurred. The youthful sun was just starting to break through the dense, steamy cloud base and Luca took advantage of this potent new energy source, developing the vital process on which most life on Earth now depends: photosynthesis, the ability to transform carbon dioxide and water into food, using sunlight. Like modern industrial systems, photosynthetic life forms – commonly known as plants – take in what they need and excrete the rest as waste. In their case, however, the 'waste' is oxygen.

For a billion years or so it didn't really matter how much oxygen our ancient photosynthetic ancestors pumped out: it was quickly absorbed by vast quantities of minerals lying around from the early formation of the Earth. This led, among other things, to the vast iron ore deposits which we mine today to make steel. For another billion years or so these minerals absorbed all the free oxygen that early organisms could churn out – and things remained pretty much unchanging and hum-drum, so far as life was concerned.

Then, about 2400 million years ago, like a garbage tip that has taken all the rubbish it can hold, the minerals reached satur-ation and, slowly, molecule by molecule, the amount of oxygen in the atmosphere began to climb. For a time, this didn't much affect early life: it continued as it always had, soaking up carbon dioxide and excreting oxygen, warmed and driven by the sun. Indeed, there was even an advantage, since the accumulating oxygen also formed ozone which served as a shield against the deadly ultraviolet rays – a sort of Planetary parasol built from the excretions of life.

After a while, however, like one of those old mediaeval cities where people threw their filth into the public streets, oxygen levels began to soar and the Planet gradually grew foul. The early life forms began, literally, to suffocate in their own mess and to die in droves. From being of little significance, the oxygen had emerged as a deadly Planet-wide poison – catastrophically so.

At this point the whole of life began to suffer what scientists have termed the 'Oxygen Holocaust' or 'Great Oxidation Event', a period of universal global poisoning, leading to the mass extinction of early life.[1] It was the worst die-off in the history of life on Earth. However, a handful of organisms survived, chiefly those that managed to adapt, over millions of years, to using oxygen in their own metabolism. Some of these, the fossil boneyard reveals, went on to become multicellular life, then fish, then animals, and the animals went on to become us.

And we, a single species, unthinkingly went on to pollute the world anew. To create a man-made chemical holocaust, not just with one chemical but with many thousands, released in billions of tonnes.

The moral of this true story?

If you foul your Planet, it can kill you.

Julian Cribb, Canberra, 2021

NOTES

Chapter 1: Chemical Avalanche

1 Hughes J, Tomoko Uemura R.I.P. The Digital Journalist, 2000. http://digitaljournalist.org/issue0007/hughes.htm

2 W Eugene Smith and Aileen M Smith, *Minamata: The Story of the Poisoning of a City, and of the People Who Choose to Carry the Burden of Courage*, Holt, Reinhart and Winston, 1975.

3 Smith and Smith, 1975, op. cit.

4 Wikipedia – W Eugene Smith. https://en.wikipedia.org/wiki/W._Eugene_Smith

5 Kazantzis G, Occupational Disease, *Encyclopaedia Britannica*, 2020. www.britannica.com/science/occupational-disease

6 These were the first occupational cancers to be described. In 1775 a surgeon, Sir Percival Pott, associated them with the occupation of chimney sweep. In the 1930s their cause was linked to exposure to polycyclic hydrocarbons, such as benzopyrene, the residues of burning wood and coal. https://en.wikipedia.org/wiki/Chimney_sweeps%27_carcinoma

7 Wikipedia, Phossy jaw, 2020. https://en.wikipedia.org/wiki/Phossy_jaw

8 UNEP, Global Chemicals Outlook II: Summary for Policymakers, March 2019. https://papersmart.unon.org/resolution/uploads/k1900123.pdf#overlay-context=pre-session-unea-4

9 US EPA, Toxic Substances Control Act (TSCA) chemical inventory, February 2020. https://www.epa.gov/tsca-inventory

10 Debunking the myths: Are there really 84,000 chemicals? ChemicalSafetyFacts.org, 2020. https://www.chemicalsafetyfacts.org/chemistry-context/debunking-myth-chemicals-testing-safety/

11 US Agency for Toxic Substances and Disease Registry (ATSDR), Chemicals, Cancer and You, 2010. www.atsdr.cdc.gov/emes/public/docs/Chemicals,%20Cancer,%20and%20You%20FS.pdf

12 European Chemicals Agency database, February 2020. www.echa.europa.eu/information-on-chemicals/ec-inventory

13 Chemical Inspection and Regulation Service (CIRS), 2016. Inventory of existing chemical substance in China. www.cirs-reach.com/news-and-articles/the-inventory-of-existing-chemical-substance-in-china-iecsc-2013-and-updates.html

14 UNEP, Global Chemicals Outlook: Towards sound management of chemicals, 2013. www.researchgate.net/publication/271524747_Global_Chemicals_Outlook_Towards_Sound_Management_of_Chemicals

15 Wang Z et al., Toward a global understanding of chemical pollution: A first comprehensive analysis of national and regional chemical inventories. *Environmental Science and Technology*, 54:5 (2020), 2575–84. https://pubs.acs.org/doi/abs/10.1021/acs.est.9b06379

16 UNEP, Global Chemicals Outlook II: Summary for policymakers, March 2019. https://papersmart.unon.org/resolution/uploads/k1900123.pdf#overlay-context=pre-session-unea-4

17 ATSDR 2010, op. cit.

18 UNEP 2019, op. cit.

19 UNEP 2019, op. cit.

20 Tran XT, Consequences of chemical warfare in Vietnam, March 2006; X.T. Tran, Agent Orange: Diseases associated with Agent Orange exposure, 25 March 2010. US Department of Veterans Affairs Office of Public Health and Environmental Hazards.

21 Geyer R et al., Production, use, and fate of all plastics ever made. *Science Advances* (AAAS), 3:7, 19 July 2017, e1700782. https://advances.sciencemag.org/content/3/7/e1700782

22 Borrelli P, Robinson DA, Fleischer LR et al., An assessment of the global impact of 21st century land use change on soil erosion. *Nature Communications*, 8 (2017), 2013. doi.org/10.1038/s41467-017-02142-7

23 Wilkinson BH and McElroy BJ, The impact of humans on continental erosion and sedimentation. *Geological Society of America Bulletin*, July 2006. doi.org/10.1130/B25899.1

24 Pimentel D, Soil erosion: A food and environmental threat. *Environment, Development and Sustainability*, 8 (2006), 119–37. doi.org/10.1007/s10668-005-1262-8

25 Chen M and Graedel TE, A half-century of global phosphorus flows, stocks, production, consumption, recycling, and environmental impacts. *Global Environmental Change*, 36 (January 2016), 139–52.

26 Diaz RJ and Rosenberg R, Spreading dead zones and consequences for marine ecosystems. *Science*, 321:5891 (2008), 926–9. https://science.sciencemag.org/content/321/5891/926

27   Angus I, Nitrogen crisis: A neglected threat to Earth's life support systems, April 2019. https://climateandcapitalism.com/2019/04/18/nitrogen-crisis-a-neglected-threat-earths-life-support-systems/

28   Rockstrom J et al., A safe operating space for humanity. *Nature*, 461 (2009), 472–5. https://www.nature.com/articles/461472a

29   World Mining Congress, World Mining Data 2019, Vienna, 2019. www.wmc.org.pl/sites/default/files/WMD%202019%20web.pdf

30   Stanford K, Red mud – addressing the problem. Aluminium Insider, November 2016. https://aluminiuminsider.com/red-mud-addressing-the-problem/

31   Earthworks, Tailings are mine waste, 2020. https://www.earthworks.org/issues/tailings/

32   Earthworks and MiningWatch Canada, Troubled waters, 2012. www.earthworks.org/publications/troubled_waters/

33   Blacksmith Institute/Pure Earth, Toxic Sites Identification Program, 2020. www.contaminatedsites.org/

34   Environmental effects of Iron Mountain. https://ca.water.usgs.gov/projects/iron_mountain/environment.html

35   Global Energy Statistical Yearbook, 2019.

36   Perera FP, Multiple threats to child health from fossil fuel combustion: Impacts of air pollution and climate change. *Environmental Health Perspectives*, 125:2 (February 2017), 141–8. www.ncbi.nlm.nih.gov/pmc/articles/PMC5289912/

37   Myllyvirta L, Quantifying the economic costs of air pollution from fossil fuels. Centre for Research on Energy and Clean Air, February 2020. https://energyandcleanair.org/publications/costs-of-air-pollution-from-fossil-fuels/

38   WHO, Air Pollution, 2020. www.who.int/health-topics/air-pollution#tab=tab_1

39   Landrigan PJ et al., Report of *The Lancet* Commission on Pollution and Health, 19 October 2017. https://www.thelancet.com/journals/lancet/article/PIIS0140-6736(17)32345-0/fulltext

40   Arms Control Association, June 2018. www.armscontrol.org/factsheets/cbwprolif; Chemical Weapons Convention, 2020. www.opcw.org/media-centre/opcw-numbers

41   Farrell M, Assassination and poisoning. In *Criminology of Poisoning Contexts*, Springer, 2020. https://link.springer.com/chapter/10.1007/978-3-030-40830-5_4

42 Stockholm International Peace Research Institute Year Book 2019. www.sipri.org/sites/default/files/2019-06/yb19_summary_eng_1 .pdf

43 Gill P, 1 million tons of contaminated radioactive water still plagues Japan nearly a decade after the Fukushima nuclear disaster. Business Insider, 7 August 2020. https://www.businessinsider.in/ science/environment/news/1-million-tons-of-contaminated-radioactive-water-in-japan-nearly-a-decade-after-the-fukushima-nuclear-disaster/articleshow/77405829.cms

44 Pasley J, Inside America's most toxic nuclear waste dump, where 56 million gallons of buried radioactive sludge are leaking into the Earth. Business Insider, 23 September 2019. https://www .businessinsider.com/hanford-nuclear-site-photos-toxic-waste-2019- 9?r=US&IR=T

45 Patterson H, How much radioactive waste is there in the world?, August 2019. https://nda.blog.gov.uk/2019/08/02/how-much-radioactive-waste-is-there-in-the-world/

46 The World Nuclear Waste Report, 2019. https:// worldnuclearwastereport.org/wp-content/themes/wnwr_theme/ content/World_Nuclear_Waste_Report_2019_exe_summary.pdf

47 Environmental and health effects of nuclear waste dumping in the Arctic. www.princeton.edu/~ota/disk1/1995/9504/950405.PDF

48 UNEP, Global Waste Management Outlook, 2015. https://www.unep .org/resources/report/global-waste-management-outlook

49 European Commission, Environment, Waste, March 2020. https://ec .europa.eu/environment/waste/index.htm

50 The source of this widely quoted figure appears to be a 1999 paper by Schmidt. However the European Community estimates its own hazardous waste output at 100 mt, making a global figure of 400 mt if anything an underestimate.

51 Crutzen PJ, The 'Anthropocene'. In *Earth System Science in the Anthropocene*, Springer, 2006. https://link.springer.com/chapter/10 .1007/3-540-26590-2_3

Chapter 2: Poisoning a Planet

1 Northeast Fisheries Science Centre, Persistent man-made chemical pollutants found in deep-sea octopods and squids, June 2008. www .nefsc.noaa.gov/press_release/2008/SciSpot/ss0810

2 NOAA, Overdose: Drugs and chemicals in fish, NWFSC/NOAA, 24 June 2009. www.fisheries.noaa.gov/feature-story/overdose-drugs-and-chemicals-fish

3 Naidu R, 2020, personal communication. Estimate updated in 2020 from Naidu R and Wong MH, Remediation of contaminated sites, CRC CARE, Salisbury, Australia, 2013.

4 Soil Protection, European Commission, 2006, P10. https://ec.europa .eu/environment/archives/soil/pdf/soillight.pdf

5 Turusov V et al., Dichlorodiphenyltrichloroethane (DDT): Ubiquity, persistence, and risks. *Environmental Health Perspectives*, 110:2 (February 2002), 125–8. doi.org/10.1289/ehp.02110125

6 Poulos A, *The Secret Life of Chemicals: A Guide to Chemicals in Our Environment and How to Protect Yourself*, 2019. https:// professoralfredpoulos.com/the-secret-life-of-chemicals/

7 Landrigan PJ et al., Pollution and global health: An agenda for prevention. *Environmental Health Perspectives*, 6 August 2018. https:// pubmed.ncbi.nlm.nih.gov/30118434/

8 UN, About Montreal Protocol, 2020. www.unenvironment.org/ ozonaction/who-we-are/about-montreal-protocol

9 Albrecht F and Parker CF, Healing the ozone layer: The Montreal Protocol and the lessons and limits of a global governance success story. In *Great Policy Successes*, Oxford University Press, 2019, pp. 304–22. www.diva-portal.org/smash/record.jsf?pid=diva2% 3A1381798&dswid=8772

10 Akimoto H, Global air quality and pollution. *Science*, 302 (2003), 1716. https://science.sciencemag.org/content/302/5651/1716.abstract

11 UNEP Centre for Clouds, Chemistry and Climate, The Asian Brown Cloud, 2002.

12 The East is grey. *The Economist*, 10 August 2013. https://www .economist.com/briefing/2013/08/10/the-east-is-grey

13 WHO, Air Pollution, 2020. www.who.int/health-topics/air-pollution#tab=tab_1

14 Ritchie H and Roser R, Air pollution, Our World in Data, 2019. https://ourworldindata.org/air-pollution#air-pollution-is-one-of-the-world-s-leading-risk-factors-for-death

15 Bolton D, Air pollution is now a global 'public health emergency', according to the World Health Organisation. *Independent*, 19 January 2016. www.independent.co.uk/environment/air-pollution-public-health-emergency-who-world-health-organisation-a6821256.html

16 Zhang X et al., The impact of exposure to air pollution on cognitive performance. *PNAS*, 11 September 2018. www.pnas.org/content/115/37/9193

17 Carrington D and Ko L, Air pollution causes 'huge' reduction in intelligence, study reveals. *Guardian*, 28 August 2018. www.theguardian.com/environment/2018/aug/27/air-pollution-causes-huge-reduction-in-intelligence-study-reveals

18 Landrigan PJ et al., *The Lancet* Commission on Pollution and Health, 19 October 2017. https://www.thelancet.com/commissions/pollution-and-health

19 Whiting K, A Chinese professor explains what air pollution does to your health. World Economic Forum, 25 June 2019. https://www.weforum.org/agenda/2019/06/chinese-professor-explains-what-air-pollution-does-to-your-health/

20 WHO, Indoor air pollution, 21 January 2020. www.who.int/news-room/q-a-detail/indoor-air-pollution

21 'Cocktail of chemicals' found in UK mothers' breast milk due to home furnishings. *Daily Telegraph*, UK, 16 July 2019. www.telegraph.co.uk/news/2019/07/16/cocktail-chemicals-found-uk-mothers-breast-milk-due-home-furnishings/

22 Ecology Centre, Dangers lurk behind that 'new-car smell', 2012. www.ecocenter.org/newsletter/2012-02/dangers-lurk-behind-new-car-smell

23 Dietz R et al., Three decades (1983–2010) of contaminant trends in East Greenland polar bears (*Ursus maritimus*). *Environment International*, 59 (2012), 485–93. www.sciencedirect.com/science/article/pii/S0160412012002024

24 Provieri F and Pirrone N, Mercury pollution in the Arctic and Antarctic regions. In *Dynamics of Mercury Pollution on Regional and Global Scales*, Springer, 2005.

25 Australian Antarctic Division, Pollution and Waste, 2020. www.antarctica.gov.au/environment/pollution-and-waste

26 Corsolini S, Industrial contaminants in Antarctic biota. *Journal of Chromatography A*, 1216:3 (September 2008), 598–612. www.researchgate.net/publication/23190034_Industrial_contaminants_in_Antarctic_biota

27 Fuoco R et al., *Persistent Organic Pollutants in the Antarctic Environment*. Scientific Committee on Antarctic Research, Cambridge, 2009. www.scar.org/publications/occasionals/POPs_in_Antarctica.pdf

28 Chu WL et al., Heavy metal pollution in Antarctica and its potential impacts on algae. *Polar Science*, June 2019. www.sciencedirect.com/science/article/abs/pii/S1873965218300926

29 Bessa F et al., Microplastics in gentoo penguins from the Antarctic region. *Science Reports*, 9 (2019), 14191. doi.org/10.1038/s41598-019-50621-2

30 Fretwell P et al., Bedmap2: Improved ice bed, surface and thickness datasets for Antarctica. *Cryosphere*, 7 (2013), 375–93.

31 Jones N, How fast and how far will sea levels rise? Yale Environment 360. https://e360.yale.edu/features/rising_waters_how_fast_and_how_far_will_sea_levels_rise

32 IMBIE team, Mass balance of the Antarctic Ice Sheet from 1992 to 2017. *Nature*, 558 (14 June 2018), 219–22. doi.org/10.1038/s41586-018-0179-y

33 Yeo B and Langley-Turnbaugh S, Trace element deposition on Mt Everest. *Soil Survey Horizons*, 51:4 (2010), 95–101. https://pdfs.semanticscholar.org/d5d5/0f002e9cfdbfa6a04c9fbdcdee2e2273e4d0.pdf

34 Uddin R and Huda NH, Arsenic poisoning in Bangladesh. *Oman Medical Journal*, 26:3 (May 2011), 207. https://www.ncbi.nlm.nih.gov/pmc/articles/PMC3191694/

35 Acharyya SK, Ground water arsenic pollution in West Bengal and Bangladesh: Role of Quaternary stratigraphy and sedimetation. Jadavpur University, India, 19 September 2018. https://crimsonpublishers.com/aaoa/fulltext/AAOA.000550.php

36 Auman H et al., PCBS, DDE, DDT, and TCDD-EQ in two species of albatross on Sand Island, Midway Atoll, North Pacific Ocean. *Environmental Toxicology and Chemistry*, 16:3 (1997), 498–504. https://agris.fao.org/agris-search/search.do?recordID=US9735769

37 Muir DCG et al., Toxaphene and other persistent organochlorine pesticides in three species of albatrosses from the North and South Pacific Ocean. *Environmental Toxicology and Chemistry*, 21:2 (2002), 413–23. https://setac.onlinelibrary.wiley.com/doi/abs/10.1002/etc.5620210226

38 Ross PC et al., Harbor seals (*Phoca vitulina*) in British Columbia, Canada, and Washington State, USA, reveal a combination of local and global polychlorinated biphenyl, dioxin, and furan signals. *Environmental Toxicology and Chemistry*, 23:1 (2004), 157–65. http://www.researchgate.net/publication/8880990

39 Burt J and Mounter B, Marine toxicologists call for chemical monitoring to be expanded on Great Barrier Reef. ABC, 20

December 2020. https://www.abc.net.au/news/2020-12-29/study-finds-thousands-of-chemicals-in-great-barrier-reef-turtles/13012558

40 European Commission, Contaminants in Seafood, 2020. https://ec .europa.eu/environment/marine/good-environmental-status/ descriptor-9/index_en.htm

41 SeaWeb, Contaminated sediments. Ocean Issue Briefs, 2013.

42 Environment Norway, Polluted marine sediments, 2020. www .environment.no/topics/marine-and-coastal-waters/hazardous-substances-in-coastal-waters/polluted-marine-sediments/

43 Gottlieb J, EPA aims to put a cap on DDT site. *Los Angeles Times*, 12 June 2009. www.latimes.com/archives/la-xpm-2009-jun-12-me-palos-verdes-pollution12-story.html

44 Burreau S et al., Biomagnification of PBDEs and PCBs in food webs from the Baltic Sea and the northern Atlantic Ocean. *Science of The Total Environment*, 366:2–3 (1 August 2006), 659–72. www.ncbi.nlm .nih.gov/pubmed/16580050

45 Romero-Romero S et al., Biomagnification of persistent organic pollutants in a deep-sea, temperate food web. *Science of The Total Environment*, 605–6 (15 December 2017), 589–97. www.sciencedirect .com/science/article/pii/S0048969717315632

46 Cao L et al., Biomagnification of methylmercury in a marine food web in Laizhou Bay (North China) and associated potential risks to public health. *Marine Pollution Bulletin*, January 2020. www .sciencedirect.com/science/article/pii/S0025326X1930918X

47 Fox AL et al., Mercury biomagnification in food webs of the northeastern Chukchi Sea, Alaskan Arctic. *Deep Sea Research*, October 2017. https://www.sciencedirect.com/science/article/abs/pii/ S0967064517301443

48 Rumbold DG, Primer on methylmercury biomagnification in the Everglades. In Rumbold D, Pollman C, Axelrad D (eds), *Mercury and the Everglades. A Synthesis and Model for Complex Ecosystem Restoration*, Springer, 2019.

49 Kang W, Methylmercury in seawater and its bioaccumulation in marine food webs of the Canadian Arctic. University of Manitoba, 2019. http://hdl.handle.net/1993/33838

50 Espejo W et al., Trophic transfer of cadmium in marine food webs from Western Chilean Patagonia and Antarctica. *Marine Pollution Bulletin*, 137 (December 2018), 246–51. https://www.sciencedirect .com/science/article/abs/pii/S0025326X18307240

51 Shilla D et al., Trophodynamics and biomagnification of trace metals in aquatic food webs: The case of Rufiji estuary in Tanzania.

Applied Geochemistry, 100 (January 2019), 160–8. doi.org/10.1016/j
.apgeochem.2018.11.016

52 Malm O, Gold mining as a source of mercury exposure in the
    Brazilian Amazon. *Environmental Research*, 77:2 (1998), 73–8. https://
    www.sciencedirect.com/science/article/abs/pii/S0013935198938282

53 Netting J, Pesticides implicated in declining frog numbers. *Nature*,
    2000. www.nature.com/articles/35048736

54 FAO, Declining bee populations pose threat to global food security
    and nutrition, 2020. www.fao.org/news/story/en/item/1194910/
    icode/

55 WWF, Solving plastic through accountability. Worldwide Fund for
    Nature, 2019. https://d2ouvy59p0dg6k.cloudfront.net/downloads/
    plastic_update_last_03_25.pdf

56 Geyer R et al., Production, use, and fate of all plastics ever made.
    Science Advances, 3:7 (19 July 2017), e1700782. https://advances
    .sciencemag.org/content/3/7/e1700782

57 Kane IA et al., Seafloor microplastic hotspots controlled by deep-sea
    circulation. *Science* (30 April 2020), eaba5899. https://science
    .sciencemag.org/content/early/2020/04/29/science.aba5899

58 Great Pacific Garbage Patch. *National Geographic*. www
    .nationalgeographic.org/encyclopedia/great-pacific-garbage-patch/

59 Wilcox C et al., Threat of plastic pollution to seabirds is global,
    pervasive, and increasing. *PNAS* (31 August 2015). https://www.pnas
    .org/content/112/38/11899

60 Paleczny M et al., Population trend of the world's monitored
    seabirds, 1950–2010. *PLoS One* (9 June 2015). doi.org/10.1371/journal
    .pone.0129342

61 Astrom L, Shedding of synthetic microfibers from textiles. Goteborg
    University, 2015. https://link.springer.com/article/10.1007/s11356-
    017-0528-7

62 Seasave, 2020. https://seasave.org/plastic-pollution/

63 Cedervall T et al., Brain damage and behavioural disorders in fish
    induced by plastic nanoparticles delivered through the food chain.
    Scientific Reports, 7:1 (2017). www.nature.com/articles/s41598-017-
    10813-0

64 Lusher A, Hollman P and Mendoza-Hill J, Microplastics in fisheries
    and aquaculture: Status of knowledge on their occurrence and
    implications for aquatic organisms and food safety. FAO
    Fisheries and Aquaculture Technical Paper, 2017. www.fao.org/3/a-
    i7677e.pdf

65 Rios L et al., Characterisation of microplastics and toxic chemicals extracted from microplastic samples from the North Pacific Gyre. *Environmental Chemistry*, 12:5 (2015), 611–17. http://dx.doi.org/10.1071/EN14236

66 Teuten E et al., Transport and release of chemicals from plastics to the environment and to wildlife. *Philosophical Transactions of the Royal Society B*, 364 (2009), 2027–45. doi.org/10.1098/rstb.2008.0284

67 Rios L et al., Quantitation of persistent organic pollutants adsorbed on plastic debris from the Northern Pacific Gyre's 'eastern garbage patch'. *Journal of Environmental Monitoring*, 12 (2010), 2226–36. www.publish.csiro.au/en/EN14236

68 Deutsche Welle (DW), 2019. Oil companies pivot to plastics to stave off losses from fuel demand. www.dw.com/en/plastic-oil-petrochemicals-coronavirus/a-52834661

69 5 Gyres Institute. www.5gyres.org/

70 Beck EC, The Love Canal tragedy, EPA, 1979. https://archive.epa.gov/epa/aboutepa/love-canal-tragedy.html

71 Verhovek SH, After 10 years, the trauma of Love Canal continues. *New York Times*, 5 August 1988. www.nytimes.com/1988/08/05/nyregion/after-10-years-the-trauma-of-love-canal-continues.html

72 Environmental Working Group, 2010, Chromium-6 in U.S. tap water. https://static.ewg.org/reports/2010/chrome6/chrome6_report_2.pdf?_ga=2.17898673.961255454.1587185238-809310208.1587185238

73 Environmental Working Group, 2020. Chromium-6. www.ewg.org/tapwater/reviewed-chromium-6.php

74 Popovich N et al., The Trump Administration is reversing 100 environmental rules. Here's the full list. *New York Times*, 20 May 2020. www.nytimes.com/interactive/2020/climate/trump-environment-rollbacks.html

75 Rott N, Trump Administration weakens auto emissions standards. NPR, 31 March 2020. www.npr.org/2020/03/31/824431240/trump-administration-weakens-auto-emissions-rolling-back-key-climate-policy

76 US EPA, EPA announces enforcement discretion policy for COVID-19 pandemic, 26 March 2020. www.epa.gov/newsreleases/epa-announces-enforcement-discretion-policy-covid-19-pandemic

77 Lavelle M et al., Trump's move to suspend enforcement of environmental laws is a lifeline to the oil industry. Inside Climate

News, 27 March 2020. https://insideclimatenews.org/news/27032020/
coronavirus-covid-19-EPA-API-environmental-enforcement

78 Nunez C, Water pollution is a rising global crisis. *National
    Geographic*, 2020. www.nationalgeographic.com/environment/
    freshwater/pollution/

79 Blaettler K, The effects of water pollution around the world.
    Sciencing, 29 July 2019. https://sciencing.com/effects-water-
    pollution-around-world-6456.html

80 7 Biggest water-polluting countries. All About Water Filters, 2020.
    https://all-about-water-filters.com/producers-water-pollution-
    around-the-world/#tab-con-2

81 Hirani P and Dimble V, Water pollution is killing millions of
    Indians. Here's how technology and reliable data can change that.
    WEF, 4 October 2019. www.weforum.org/agenda/2019/10/water-
    pollution-in-india-data-tech-solution/

82 Dalin C et al., Groundwater depletion embedded in international
    food trade. *Nature*, 543 (30 March 2017). www.nature.com/articles/
    nature21403

83 Groundwater.org. Groundwater contamination, 2020. www
    .groundwater.org/get-informed/groundwater/contamination.html

84 World Resources Institute, Interactive Map of Eutrophication &
    Hypoxia, 2020. www.wri.org/our-work/project/eutrophication-and-
    hypoxia/interactive-map-eutrophication-hypoxia

85 Breitberg D et al., Declining oxygen in the global ocean and coastal
    waters. *Science*, 359:6371 (5 January 2018). https://science
    .sciencemag.org/content/359/6371/eaam7240

86 Mississippi Dead Zone, NASA, 2007. www.nasa.gov/vision/earth/
    environment/dead_zone.html

87 UNEP, Global Chemicals Outlook, 2012.

88 UNEP, Global Chemicals Outlook II, 2019. https://wedocs.unep.org/
    bitstream/handle/20.500.11822/28187/GCO-II_Intro.pdf?sequence=
    1&isAllowed=y

Chapter 3: Are You a Contaminated Site?

1 Roy R and Agarwal V, Tainted school lunch kills at least 22 Indian
   children. *Wall Street Journal*, 18 July 2013. http://online.wsj.com/
   article/SB10001424127887323993804578611272207737576.html

2 Carson R, *Silent Spring*. Houghton Mifflin, 1962. https://rachelcarson
   .org/SilentSpring.aspx

3 CDC, Fourth National Report on Human Exposure to Environmental Chemicals, 2019. www.cdc.gov/exposurereport/pdf/FourthReport_UpdatedTables_Volume1_Jan2019-508.pdf

4 CDC, Chemicals in the Fourth Report: Updated Tables, January 2019. www.cdc.gov/exposurereport/pdf/Report_Chemical_List-508.pdf

5 Crinnion W, The CDC fourth national report on human exposure to environmental chemicals: What it tells us about our toxic burden and how it assists environmental medicine physicians. *Alternative Medicine Review*, 15:2 (July 2010), 101–9. https://pubmed.ncbi.nlm.nih.gov/20806995/

6 CDC, Frequently asked questions. www.cdc.gov/exposurereport/faq.html

7 CDC, Chemical factsheets, 2020. www.cdc.gov/biomonitoring/chemical_factsheets.html

8 Duncan DE, Chemicals within us: My journalist-as-guinea-pig experiment is taking a disturbing turn. *National Geographic*, 2020. www.nationalgeographic.com/science/health-and-human-body/human-body/chemicals-within-us/

9 Statista, Consumption of chemicals in selected countries in 2018, 2020. www.statista.com/statistics/272287/consumption-of-chemicals-by-country-2008/

10 UK House of Commons Environmental Audit Committee, Toxic chemicals in everyday life, July 2019. https://publications.parliament.uk/pa/cm201719/cmselect/cmenvaud/1805/1805.pdf

11 WHO, Assessment of capacity in WHO EURO member states to address health-related aspects of chemical safety, 2012. https://www.euro.who.int/en/health-topics/environment-and-health/health-impact-assessment/publications/2012/assessment-of-capacity-in-who-euro-member-states-to-address-health-related-aspects-of-chemical-safety

12 WHO, Dioxins and their effects on human health, 4 October 2016. www.who.int/en/news-room/fact-sheets/detail/dioxins-and-their-effects-on-human-health

13 Environmental Working Group, Toxic chemicals found in minority cord blood, 2009. www.ewg.org/news/news-releases/2009/12/02/toxic-chemicals-found-minority-cord-blood

14 Environmental Working Group, Pollution in people: Cord blood contaminants in minority newborns, 2009. http://static.ewg.org/reports/2009/minority_cord_blood/2009-Minority-Cord-Blood-Report.pdf

15 G7–8, 1997 Declaration of the Environment Leaders of the Eight on
    Children's Environmental Health, Miami, Florida, 5–6 May 1997.
    www.g8.utoronto.ca/environment/1997miami/children.html

16 The Environment and Child Health International Birth Cohort
    Group. www.ncbi.nlm.nih.gov/pubmed/31327570

17 See Baldacci et al., 2018; Barbone et al., 2019; Botton et al., 2016;
    Boucher et al., 2009; Casas et al., 2015; Clemente et al., 2016;
    Dzwilewski and Schantz, 2015; Hertz-Picciotto et al., 2008;
    Huang et al., 2016; Iszatt et al., 2015; Perera et al., 2006; Pilsner
    et al., 2009; Rauh et al., 2011; Shelton et al., 2014; Trasande
    et al., 2009.

18 Suk W et al., Environmental pollution: An under-recognized threat
    to children's health, especially in low- and middle-income
    countries. *Environmental Health Perspectives*, 124:3 (1 March 2016). doi
    .org/10.1289/ehp.1510517

19 Weldon RH et al., A pilot study of pesticides and PCBs in the breast
    milk of women residing in urban and agricultural communities of
    California. *Journal of Environmental Monitoring*, 13:11 (2011), 3136–44.
    https://pubmed.ncbi.nlm.nih.gov/22009134/

20 Massart F et al., Chemical biomarkers of human breast milk
    pollution. Biomark Insights, 2008. www.ncbi.nlm.nih.gov/pmc/
    articles/PMC2688366/

21 Qu W et al., Exposure of young mothers and newborns to
    organochlorine pesticides (OCPs) in Guangzhou, China. *Science of The
    Total Environment*, 408:16 (July 2010), 3133–8. pubmed.ncbi.nlm.nih
    .gov/20471063/

22 Bedi JS et al., Pesticide residues in human breast milk: Risk
    assessment for infants from Punjab, India. *Science of The Total
    Environment*, 463–4 (October 2013), 720–61. www.ncbi.nlm.nih.gov/
    pubmed/23850662

23 European Environment and Health Information Service (ENHIS),
    Persistent organic pollutants in human milk, Fact Sheet 4.3, 2009.
    www.euro.who.int/__data/assets/pdf_file/0003/97032/4.3.-Persistant-
    Organic-Pollutantsm-EDITED_layouted_V2.pdf?ua=1

24 International Pollutants Elimination Network, PFAS pollution
    across the Middle East and Asia, 2019. www.documentcloud.org/
    documents/5980916-Pfas-Pollution-Across-the-Middle-East-and-Asia
    .html

25 Glyphosate found in breast milk. *The Ecologist*, 28 April 2014. https://
    theecologist.org/2014/apr/28/glyphosate-found-breast-milk

26 'Cocktail of chemicals' found in UK mothers' breast milk due to
    home furnishings. *Daily Telegraph*, UK, 16 July 2019. www.telegraph
    .co.uk/news/2019/07/16/cocktail-chemicals-found-uk-mothers-
    breast-milk-due-home-furnishings/

27 Fang J et al., Spatial and temporal trends of the Stockholm
    Convention POPs in mothers' milk: A global review. *Environmental
    Science and Pollution Research*, 22 (2015), 8989–9041. https://pubmed
    .ncbi.nlm.nih.gov/25913228/

28 Van den Berg M et al., WHO/UNEP global surveys of PCDDs, PCDFs,
    PCBs and DDTs in human milk and benefit–risk evaluation of
    breastfeeding. *Archives of Toxicology*, 91:1 (2017), 83–96. https://
    pubmed.ncbi.nlm.nih.gov/27438348/

29 Mead N, Contaminants in human milk: Weighing the risks against
    the benefits of breastfeeding. *Environmental Health Perspectives*,
    116:10 (October 2008), A426–34.

30 Mellowship D, *Toxic Beauty*. Octopus, 2010.

31 Statista, Beauty and personal care market value worldwide from
    2018 to 2025. www.statista.com/statistics/550657/beauty-market-
    value-growth-worldwide-by-country/

32 Toronto University, Guidelines on the use of perfumes and scented
    products. 2020. https://ehs.utoronto.ca/our-services/occupational-
    hygiene-safety/guidelines-on-the-use-of-perfumes-and-scented-
    products/

33 Environmental Working Group, Skin deep, 2020. www.ewg.org/
    skindeep/contents/about-page/

34 Environmental Working Group, The toxic twenty chemicals and
    contaminants in cosmetics, 2019. https://cdn3.ewg.org/sites/default/
    files/u352/Toxic%20Twenty%20Report_2.pdf?_ga=2.193122989
    .1257270375.1588045010-809310208.1587185238

35 Campaign for Safe Cosmetics, Chemicals of concern, 2020. www
    .safecosmetics.org/get-the-facts/chem-of-concern/

36 Xu S et al., Adverse events reported to the US Food and Drug
    Administration for cosmetics and personal care products. *JAMA
    Internal Medicine*, 177:8 (2017), 1202–4. https://jamanetwork.com/
    journals/jamainternalmedicine/fullarticle/2633256

37 US FDA, Prohibited & restricted ingredients in cosmetics, 2020.
    www.fda.gov/cosmetics/cosmetics-laws-regulations/prohibited-
    restricted-ingredients-cosmetics

38 List of substances prohibited in cosmetic products. https://ec.europa
    .eu/growth/tools-databases/cosing/pdf/COSING_Annex%20II_v2.pdf

39   CHOICE, Chemicals in cosmetics, March 2016. www.choice.com.au/
     health-and-body/beauty-and-personal-care/skin-care-and-cosmetics/
     articles/chemicals-in-cosmetics

40   Taking the 'real food' challenge'. RN First Bite, Radio National.
     www.abc.net.au/radionational/programs/rnfirstbite/taking-the-
     27real-food27-challenge/5041294

41   Leake L, 100 Days of Real Food, 2020. www.100daysofrealfood.com/

42   WHO, Assessing chemical risks in food. www.who.int/activities/
     assessing-chemical-risks-in-food

43   WHO, Pesticide residues in food, 2020. https://www.who.int/news-
     room/fact-sheets/detail/pesticide-residues-in-food

44   Worldometer, 2020. www.worldometers.info/food-agriculture/
     pesticides-by-country/

45   Tang FHM et al, Risk of pesticide pollution at the global scale, *Nature*, 29
     Mar 2021. https://protect-eu.mimecast.com/s/63FpCx6x6tPKZnVc85BBu?
     domain=nature.com https://www.nature.com/articles/s41561-021-00712-5

46   Matysiak M et al., Effect of prenatal exposure to pesticides on
     children's health. *Journal of Environmental Pathology, Toxicology and
     Oncology*, July 2016. www.dl.begellhouse.com/journals/
     0ff459a57a4c08d0,330e808854c98f7b,43e22c944604ab3b.html

47   Napier A, Salmon and Trout Conservation argue with Alresford
     Salads over pollution in River Itchen. *Hampshire Chronicle*, 2020.
     www.hampshirechronicle.co.uk/news/18602214.salmon-trout-
     conservation-argue-alresford-salads/

48   European Commission, Food Safety: Food Additives. https://webgate
     .ec.europa.eu/foods_system/main/?sector=FAD&auth=SANCAS

49   Kobylewski S, Food dyes: A rainbow of risks. Center for Science in
     the Public Interest, 2010. https://cspinet.org/sites/default/files/
     attachment/food-dyes-rainbow-of-risks.pdf

50   Anand SP and Sati N, Artificial preservatives and their harmful
     effects: Looking toward nature for safer alternatives. *International
     Journal of Pharmaceutical Sciences and Research*, 1 July 2013. https://ijpsr
     .com/bft-article/artificial-preservatives-and-their-harmful-effects-
     looking-toward-nature-for-safer-alternatives/?view=fulltext

51   Preservatives, ChemicalSafetyFacts.org. www.chemicalsafetyfacts
     .org/preservatives/#answering-questions

52   Munke J et al., Impacts of food contact chemicals on human health:
     A consensus statement. *Environmental Health*, 25, 2020. https://
     ehjournal.biomedcentral.com/articles/10.1186/s12940-020-0572-5

53   US NIH, Food Additives, Contaminants, Carcinogens, and Mutagens,
     1983. www.ncbi.nlm.nih.gov/books/NBK216714/

54 McCarthy C, Common food additives and chemicals harmful to children. *Harvard Health*, 24 June 2018. www.health.harvard.edu/blog/common-food-additives-and-chemicals-harmful-to-children-2018072414326

55 Trasande L et al., Food additives and child health. *Pediatrics*, 142:2 (August 2018), e20181408. https://pediatrics.aappublications.org/content/142/2/e20181408

56 Erickson MC and Doyle MP, The challenges of eliminating or substituting antimicrobial preservatives in foods. *Annual Reviews of Food Science and Technology*, 2017. www.annualreviews.org/doi/full/10.1146/annurev-food-030216-025952

57 Organic foods market 2020 global industry revenue by applications, market share, driving forces, competitive situation forecast to 2025. Absolute Reports, March 2020. www.globenewswire.com/news-release/2020/03/02/1993359/0/en/Organic-Foods-Market-2020-Global-Industry-Revenue-by-Applications-Market-Share-Driving-Forces-Competitive-Situation-Forecast-to-2025.html

58 Environmental Working Group, Clean fifteen, 2020. www.ewg.org/foodnews/clean-fifteen.php

59 Environmental Working Group, These are the foods with the most pesticides, 2020. www.eatthis.com/foods-most-pesticides/

60 EU pesticides database, 2020. https://ec.europa.eu/food/plant/pesticides/eu-pesticides-database/public/?event=activesubstance.selection&language=EN

61 The 2018 European Union report on pesticide residues in food. *European Food Safety Authority Journal*, 2020. https://efsa.onlinelibrary.wiley.com/doi/full/10.2903/j.efsa.2020.6057

62 EFSA, Cumulative risk assessment of pesticides: FAQ. www.efsa.europa.eu/en/news/cumulative-risk-assessment-pesticides-faq

63 Wikipedia, Food safety incidents in China. https://en.wikipedia.org/wiki/Food_safety_incidents_in_China

64 Wikipedia, 2008 Chinese milk scandal. https://en.wikipedia.org/wiki/2008_Chinese_milk_scandal

65 Survey shows Chinese public not satisfied with the nation's food safety. DJS Research, 25 July 2014. www.djsresearch.co.uk/FoodMarketResearchInsightsAndFindings/article/Survey-Shows-Chinese-Public-Not-Satisfied-with-the-Nations-Food-Safety-01621

66 FleishmanHillard, Food safety in Asia, 3 May 2016. https://fleishmanhillard.com/2016/05/food-agriculture-beverage/food-safety-in-asia/

67 Wikipedia, List of food contamination incidents. https://en.wikipedia.org/wiki/List_of_food_contamination_incidents

68   Cribb JHJ, *Food or War*, Cambridge University Press, 2019.

69   Food Intolerance Network 2020. http://fedup.com.au

70   WHO, Chemical hazards in drinking-water, 2020. www.who.int/
water_sanitation_health/water-quality/guidelines/chemicals/en/

71   The Business Research Company, The global bottled water
market. 28 February 2018. https://blog.marketresearch.com/the-
global-bottled-water-market-expert-insights-statistics

72   Wagner M et al., Identification of putative steroid receptor
antagonists in bottled water: Combining bioassays and high-
resolution mass spectrometry. *PLoS One*, 28 August 2013. https://
journals.plos.org/plosone/article?id=10.1371/journal.pone
.0072472

73   Metz CM, Bisphenol A: Understanding the controversy. *Workplace
Health and Safety*, 22 January 2016. https://journals.sagepub.com/doi/
10.1177/2165079915623790

74   George L, Avoid plastic poisoning. MrWaterGeek, 2020. www
.mrwatergeek.com/check-bottled-water-for-bpa/#LDPE

75   Mekonnen M and Hoekstra AY, Four billion people facing severe
water scarcity. *Science Advances*, 2:2 (12 February 2016),
e1500323. https://advances.sciencemag.org/content/2/2/
e1500323?utm_source=TrendMD&utm_medium=cpc&utm_
campaign=TrendMD_0

76   US Director of National Intelligence, Global Water Security, 2012.
www.dni.gov/files/documents/Special%20Report_ICA%20Global%
20Water%20Security.pdf

77   UNEP, The snapshot report of the world's water quality, 2016. www
.unenvironment.org/resources/publication/snapshot-report-worlds-
water-quality

78   World Bank, China: Agenda for water sector strategy for North
China, 9 May 2002. http://documents.worldbank.org/curated/en/
914501468770501750/pdf/multi0page.pdf

79   Xia C, 80% underground water undrinkable in China. China.org.cn,
11 April 2016. www.china.org.cn/environment/2016-04/11/content_
38218704.htm

80   Gibson C, Water pollution in China is the country's worst
environmental issue. The Borgen Project, 10 March 2018. https://
borgenproject.org/water-pollution-in-china/

81   Buckley C and Piao V, Rural water, not city smog, may be China's
pollution nightmare. *New York Times*, 9 April 2016. www.nytimes
.com/2016/04/12/world/asia/china-underground-water-pollution
.html

82 Hirani P et al., Water pollution is killing millions of Indians. World Economic Forum, 4 October 2019. www.weforum.org/agenda/2019/10/water-pollution-in-india-data-tech-solution/

83 European Environmental Agency, European waters: Assessment of status and pressures, 2018. www.eea.europa.eu/publications/state-of-water

84 Natural Resources Defence Council, Water pollution. NRDC, 14 May 2018. www.nrdc.org/stories/water-pollution-everything-you-need-know#whatis

85 Environmental Working Group, Mapping the PFAS contamination crisis: New data show 1,582 sites in 49 states, 2020. www.ewg.org/interactive-maps/pfas_contamination/

86 Environmental Working Group, Chromium-6 in U.S. tap water, 2010. https://static.ewg.org/reports/2010/chrome6/chrome6_report_2.pdf?_ga=2.17898673.961255454.1587185238-809310208.1587185238

87 Andrews DQ et al., Population-wide exposure to per- and polyfluoroalkyl substances from drinking water in the United States. *Environmental Science and Technology Letters*, 7:12 (2020), 931–6. https://pubs.acs.org/doi/10.1021/acs.estlett.0c00713

88 Denchak M, Flint water crisis: Everything you need to know. NRDC, 8 November 2018. www.nrdc.org/stories/flint-water-crisis-everything-you-need-know#sec-summary

89 Oram B, Disinfection byproducts: Trihalomethanes. Water Research Centre, 2020. https://water-research.net/index.php/trihalomethanes-disinfection

90 Waller K et al., Trihalomethanes in drinking water and spontaneous abortion. *Epidemiology*, 9:2 (March 1998), 134–40. www.jstor.org/stable/3702950?seq=1#metadata_info_tab_contents

91 Mullee A et al., Association between soft drink consumption and mortality in 10 European countries. *JAMA Internal Medicine*, 2019. https://jamanetwork.com/journals/jamainternalmedicine/article-abstract/2749350

92 Harvard School of Public Health, Sugary drinks, 2020. www.hsph.harvard.edu/nutritionsource/healthy-drinks/sugary-drinks/

93 Kregiel D, Health safety of soft drinks: Contents, containers, and microorganisms. *Biomed Research International*, 2015. www.ncbi.nlm.nih.gov/pmc/articles/PMC4324883/

94 Hensrud D, Does coffee offer health benefits? Mayo Clinic, 2020. www.mayoclinic.org/healthy-lifestyle/nutrition-and-healthy-eating/expert-answers/coffee-and-health/faq-20058339

95 Older Adults and Pesticides, Fact Sheet. http://npic.orst.edu/
factsheets/olderadults.pdf

96 Dent B, The hydrogeological context of cemetery operations and
planning in Australia. University of Technology Sydney, Sydney,
2002. https://opus.lib.uts.edu.au/handle/2100/963

97 Landrigan PJ et al., *The Lancet* Commission on pollution and health. *The
Lancet* Commissions, 391:10119 (3 February 2018), 462–512. www
.thelancet.com/journals/lancet/article/PIIS0140-6736(17)32345-0/fulltext

98 Broadfoot M, Pollution is a global but solvable threat to health, say
scientists. Environmental Factor, January 2019. https://factor.niehs
.nih.gov/2019/1/science-highlights/pollution/index.htm

99 WHO, Cancer, 2020. www.who.int/health-topics/cancer#tab=tab_1

Chapter 4: Diabolic Cocktail

1 Frienkel S, Trace chemicals in everyday food packaging cause worry
over cumulative threat. *The Washington Post*, 17 April 2012. www
.washingtonpost.com/national/health-science/trace-chemicals-in-
everydayfood-packaging-cause-worry-over-cumulativethreat/2012/
04/16/gIQAUILvMT_story.html

2 Rudel RA et al., Food packaging and bisphenol A and bis(2-
ethyhexyl) phthalate exposure: Findings from a dietary
intervention. *Environmental Health Perspectives*, 1 July 2011. www.ncbi
.nlm.nih.gov/pmc/articles/PMC3223004/?tool=pubmed

3 US National Research Council, *Science and Decisions: Advancing Risk
Assessment*. National Academies Press, 2009. www.nap.edu/catalog/
12209/science-and-decisions-advancing-risk-assessment

4 ChemicalSafetyFacts.org, The dose makes the poison, 2020. www
.chemicalsafetyfacts.org/wp-content/uploads/2016/12/the-dose-
makes-the-poison.jpg

5 US EPA, Guidelines for the health risk assessment of chemical
mixtures, September 1986. www.epa.gov/sites/production/files/
2014-11/documents/chem_mix_1986.pdf

6 Drakvik E et al., Statement on advancing the assessment of
chemical mixtures and their risks for human health and the
environment. *Environment International*, 134 (January 2020), 105267.
www.sciencedirect.com/science/article/pii/S0160412019331538#!

7 Henn BC, Chemical mixtures and children's health. *Current Opinion
in Pediatrics*, 26 :2 (April 2014), 223–9. www.ncbi.nlm.nih.gov/pmc/
articles/PMC4043217/

8 Christiansen S et al., Synergistic disruption of external male sex organ development by a mixture of four antiandrogens. *Environmental Health Perspectives*, 117:12 (2009), 1839–46. https://ehp.niehs.nih.gov/doi/full/10.1289/ehp.0900689

9 Goodson WH et al., Assessing the carcinogenic potential of low-dose exposures to chemical mixtures in the environment: The challenge ahead. *Carcinogenesis*, June 2015. www.ncbi.nlm.nih.gov/pubmed/26106142

10 Braun JM et al., Gestational exposure to endocrine-disrupting chemicals and reciprocal social, repetitive, and stereotypic behaviors in 4- and 5-year-old children: The HOME study. *Environmental Health Perspectives*, 122:5 (May 2014), 513–20. www.ncbi.nlm.nih.gov/pubmed/24622245

11 Liew Z et al., Maternal plasma perfluoroalkyl substances and miscarriage: A nested case-control study in the Danish National Birth Cohort. *Environmental Health Perspectives*, 22 April 2020, CID: 047007. https://doi.org/10.1289/EHP6202

12 Trasande L et al., Burden of disease and costs of exposure to endocrine disrupting chemicals in the European Union: An updated analysis. *Andrology*, 4:4 (July 2016), 565–72. www.ncbi.nlm.nih.gov/pubmed/27003928?dopt=Abstract

13 Kalkbrenner AE et al., Environmental chemical exposures and autism spectrum disorders: A review of the epidemiological evidence. *Current Problems in Pediatric and Adolescent Health Care*, November 2014. www.sciencedirect.com/science/article/abs/pii/S1538544214000741

14 Henn B et al., Chemical mixtures and children's health. *Current Opinions in Pediatrics*, April 2014. www.ncbi.nlm.nih.gov/pmc/articles/PMC4043217/

15 Valeri L et al., The joint effect of prenatal exposure to metal mixtures on neurodevelopmental outcomes at 20–40 months of age: Evidence from rural Bangladesh. *Environmental Health Perspectives*, 26 June 2017. www.ncbi.nlm.nih.gov/pubmed/28669934

16 Hu Z, Assessing the carcinogenic potential of low-dose exposures to chemical mixtures in the environment: Focus on the cancer hallmark of tumor angiogenesis. *Carcinogenesis*, 36, June 2015. https://academic.oup.com/carcin/article/36/Suppl_1/S184/315846

17 White AJ et al., Metallic air pollutants and breast cancer risk in a nationwide cohort study. *Epidemiology*, January 2020. www.ncbi.nlm.nih.gov/pmc/articles/PMC6269205/

18 Russo PN et al., Occupational exposures to carcinogens in an industry identifying and communicating chemical risk. *International Journal of Pollution*, 2019. www.gavinpublishers.com/articles/Review-Article/International-Journal-of-Pollution-Research/occupational-exposures-to-carcinogens-in-an-industry-identifying-and-communicating-chemical-risk

19 King A, Chemical mixtures pose 'underestimated' risk to human health say scientists. *Horizon*, 15 May 2019. https://horizon-magazine.eu/article/chemical-mixtures-pose-underestimated-risk-human-health-say-scientists.html

20 Kortencamp A et al., *State of the Art Report on Mixture Toxicity*, EU, The Hague, 2009. http://ec.europa.eu/environment/chemicals/pdf/report_Mixture%20toxicity.pdf

21 Carpenter DO et al., Understanding the human health effects of chemical mixtures. *Environmental Health Perspectives*, 100 (2002), 259–69.

22 UNEP, Global Chemicals Outlook, 2013, p. 48. https://sustainabledevelopment.un.org/content/documents/1966Global%20Chemical.pdf

23 Kortencamp A et al. 2009, op. cit.

24 McCarty LS and Borget CJ, Review of the toxicity of chemical mixtures: Theory, policy, and regulatory practice. *Regulatory Toxicology and Pharmacology*, 45:2 (July 2006), 119–43. www.sciencedirect.com/science/article/abs/pii/S0273230006000626

25 Kortencamp A et al. 2009, op. cit.

26 Kar S et al., Exploration of computational approaches to predict the toxicity of chemical mixtures. *Toxics*, 7:1 (2019), 15. www.mdpi.com/2305-6304/7/1/15/htm

27 Agency for Toxic Substances and Disease Registry, 2020. www.atsdr.cdc.gov/mixtures/index.html

28 Carpenter DO et al., Understanding the human health effects of chemical mixtures. *Environmental Health Perspectives*, 2002. https://ehp.niehs.nih.gov/doi/abs/10.1289/ehp.02110s125

29 Hope M, How the tobacco and fossil fuel industries fund disinformation campaigns around the world. Resilience, 20 February 2019. www.resilience.org/stories/2019-02-20/revealed-how-the-tobacco-and-fossil-fuel-industries-fund-disinformation-campaigns-around-the-world/

30 E.g. CDC, Environmental Chemicals, April 2017. www.cdc.gov/biomonitoring/environmental_chemicals.html

31 Wikipedia, Environmental chemistry. https://en.wikipedia.org/wiki/Environmental_chemistry

Chapter 5: Unseen Risks

1 Toxic town, *South China Morning Post*, 7 June 2005.
2 Pinghui Z, China's most notorious e-waste dumping ground now cleaner but poorer. *South China Morning Post*, 22 September 2017. www.scmp.com/news/china/society/article/2112226/chinas-most-notorious-e-waste-dumping-ground-now-cleaner-poorer
3 Staub C, China moves closer to complete import ban. Resource Recycling, 5 May 2020. https://resource-recycling.com/recycling/2020/05/05/china-moves-closer-to-complete-import-ban/
4 Pricop L, Thailand bans the import of e-waste. Inhabitat, 17 August 2018. https://inhabitat.com/thailand-bans-the-import-of-e-waste/
5 Carroll C, High-tech trash. *National Geographic*, January 2008. http://ngm.nationalgeographic.com/2008/01/high-tech-trash/carroll-text
6 UN News Centre, UN environment chief warns of 'tsunami' of e-waste at conference on chemical treaties. UNSDG, 5 May 2015. www.un.org/sustainabledevelopment/blog/2015/05/un-environment-chief-warns-of-tsunami-of-e-waste-at-conference-on-chemical-treaties/
7 Tonnes of electronic waste thrown out, The World Counts, 2020. www.theworldcounts.com/challenges/Planet-earth/waste/electronic-waste-facts
8 Ryder G et al., The world's e-waste is a huge problem. It's also a golden opportunity. World Economic Forum, 24 January 2019. www.weforum.org/agenda/2019/01/how-a-circular-approach-can-turn-e-waste-into-a-golden-opportunity/
9 Wong MH et al., *Adverse Environmental and Health Impacts of Uncontrolled Recycling and Disposal of Electronic Waste*, CRC CARE, Salisbury, Australia, 2013.
10 Vidal J, Toxic e-waste dumped in poor nations, says United Nations. UN University, 16 December 2013. https://ourworld.unu.edu/en/toxic-e-waste-dumped-in-poor-nations-says-united-nations
11 Ryder et al. 2019, op. cit.
12 The Mesothelioma Centre, Mesothelioma & Asbestos Worldwide, 2020. www.asbestos.com/mesothelioma/worldwide/
13 Apple Recycling Programs, 2020. www.apple.com/recycling/nationalservices/
14 https://support.google.com/store/answer/2664771?hl=en-AU
15 www.samsungrecycle.co.uk/
16 Nanodatabase, 2020. http://nanodb.dk/

17 Seaton A et al., Nanoparticles, human health hazard and regulation. *Journal of the Royal Society Interface*, 6 February 2010. https://doi.org/10.1098/rsif.2009.0252.focus

18 Buzea C et al., Nanomaterials and nanoparticles: Sources and toxicity. *Biointerphases*, 2 (2007), MR17–71. https://link.springer.com/article/10.1116/1.2815690

19 Baranowska-Wójcik, E et al. Effects of titanium dioxide nanoparticles exposure on human healtha review. *Biological Trace Element Research*, 193 (2020). https://doi.org/10.1007/s12011-019-01706-6

20 EFSA, Presence of microplastics and nanoplastics in food, with particular focus on seafood, 11 May 2016. https://efsa.onlinelibrary.wiley.com/doi/pdf/10.2903/j.efsa.2016.4501

21 Toussaint B et al., Review of micro- and nanoplastic contamination in the food chain. *Food Additives and Contaminants*, 15 April 2019. www.tandfonline.com/doi/full/10.1080/19440049.2019.1583381

22 Examples are ICON, https://web.archive.org/web/20070314002422/http://icon.rice.edu/ and PEN, www.wilsoncenter.org/publication-series/project-emerging-nanotechnologies

23 Hansen SF, React now regarding nanomaterial regulation. *Nature Nanotechnology*, 4 August 2017. www.nature.com/articles/nnano.2017.163

24 SafeNano, About IOM, 2020. www.safenano.org/about/aboutiom/

25 Gorini G et al., Smoking history is an important risk factor for severe COVID-19. Tobacco Control, April 2020. https://blogs.bmj.com/tc/2020/04/05/smoking-history-is-an-important-risk-factor-for-severe-covid-19/

26 World Resources Institute, Interactive map of eutrophication & hypoxia, 2020. www.wri.org/our-work/project/eutrophication-and-hypoxia/interactive-map-eutrophication-hypoxia

27 Mallapur C, 70% of urban India's sewage is untreated. IndiaSpend, 27 January 2016. https://archive.indiaspend.com/cover-story/70-of-urban-indias-sewage-is-untreated-54844

28 Breitburg D et al., Declining oxygen in the global ocean and coastal waters. *Science*, 359:6371 (5 January 2018). https://science.sciencemag.org/content/359/6371/eaam7240.abstract

29 NASA, Mississippi Dead Zone, 2020. www.nasa.gov/vision/earth/environment/dead_zone.html

30 Queensland Government, Scientific Consensus Statement, 2017. Land use impacts on Great Barrier Reef water quality and ecosystem

condition. www.reefplan.qld.gov.au/science-and-research/the-scientific-consensus-statement

31 Strokal M et al., Alarming nutrient pollution of Chinese rivers as a result of agricultural transitions. *Environmental Research Letters*, 11 (2016).

32 Murray CJ et al., Past, present and future eutrophication status of the Baltic Sea. *Frontiers in Marine Science*, 25 January 2019. www.frontiersin.org/articles/10.3389/fmars.2019.00002/full

33 Karydis M and Kitsiou D, Eutrophication and environmental policy in the Mediterranean Sea: A review. *Environmental Monitoring and Assessment*, 184 (28 September 2011). https://link.springer.com/article/10.1007/s10661–011-2313-2

34 Staletovich J, Dead fish, birds, manatees, even a whale shark. Toll from worst red tide in decade grows. *Miami Herald*, 17 August 2018. www.miamiherald.com/news/local/environment/article215839815.html

35 Scavia D, How your diet contributes to nutrient pollution and dead zones in lakes and bays, 12 July 2019. https://theconversation.com/how-your-diet-contributes-to-nutrient-pollution-and-dead-zones-in-lakes-and-bays-118902

36 UN Environment Programme, Paralysed by growth: A lake under siege. UNEP, 10 April 2018. www.unenvironment.org/news-and-stories/story/paralysed-growth-lake-under-siege

37 INI commits to support a global goal to halve nitrogen waste by 2030 with the support of the INMS project. INI, 30 October 2018. http://inms.international/news/ini-commits-support-global-goal-halve-nitrogen-waste-2030-support-inms-project

38 The Nutrient Challenge, Global Partnership on Nutrient Management (GPNM), 2020. www.nutrientchallenge.org/

39 Stockholm Resilience Centre, The nine Planetary boundaries, 2020. www.stockholmresilience.org/research/Planetary-boundaries/Planetary-boundaries/about-the-research/the-nine-Planetary-boundaries.html

40 Cribb JHJ, *Food or War*, Cambridge University Press, 2019.

41 Pesticide consumption, Our World in Data, 2020. https://ourworldindata.org/pesticides#pesticide-production

42 Estimates of food production losses due to diseases, insects and weeds. Our World in Data, 2020. https://ourworldindata.org/pesticides#pesticide-production

43 The neonics include acetamiprid, clothianidin, imidacloprid, nitenpyram, nithiazine, thiacloprid and thiamethoxam.

44 A typical example is Woodcock BA et al., Country-specific effects of neonicotinoid pesticides on honey bees and wild bees. *Science*, 356:6345) (30 June 2017), 1393–5. https://science.sciencemag.org/content/356/6345/1393.abstract

45 FAO, FAO's global action on pollination services for sustainable agriculture, 2019. www.fao.org/pollination/en/

46 Van Klink R et al., Meta-analysis reveals declines in terrestrial but increases in freshwater insect abundances. *Science*, 368:6489) (24 April 2020), 417–20. https://science.sciencemag.org/content/368/6489/417.abstract

47 Borenstein S, Earth's insect population shrunk 'jaw-dropping' 27% over past 30 years. AP/*Time*, 23 April 2020. https://time.com/5826457/insects-global-decline/

48 European Commission, Neonicotinoids. Current status of the neonicotinoids in the EU, 2020. https://ec.europa.eu/food/plant/pesticides/approval_active_substances/approval_renewal/neonicotinoids_en

49 US EPA, Proposed interim registration review decision for neonicotinoids, 22 January 2020. www.epa.gov/pollinator-protection/proposed-interim-registration-review-decision-neonicotinoids

50 Sass J, NRDC briefs Congress on neonic pesticide human health harms, 15 October 2019. www.nrdc.org/experts/jennifer-sass/nrdc-briefs-congress-neonic-pesticide-human-health-harms

51 Flores G, A political battle over pesticides. *The Scientist*, 10 April 2013. www.the-scientist.com/?articles.view/articleNo/35058/title/A-Political-Battle-Over-Pesticides

52 UN report warns of consequences of chemical intensification. International Chemical Secretariat, 27 September 2012. https://chemsec.org/un-report-warns-of-consequences-of-chemical-intensification/

53 UNEP Global Chemicals Outlook, 2012. www.researchgate.net/publication/271524747_Global_Chemicals_Outlook_Towards_Sound_Management_of_Chemicals

54 UNEP Global Chemical Outlook II, 29 April 2019. www.unenvironment.org/explore-topics/chemicals-waste/what-we-do/policy-and-governance/global-chemicals-outlook

55 Xinhua, 11 March 2019. www.xinhuanet.com/english/2019-03/11/c_137886702.htm

56 OPCW, OPCW by the numbers, 2020. www.opcw.org/media-centre/
    opcw-numbers

57 Bull JMR, The deadliness below. *DailyPress*, 3 October 2005. www
    .dailypress.com/news/dp-02761sy0oct30-story.html

58 Bearden DM, U.S. disposal of chemical weapons in the ocean:
    Background and issues for Congress. CRS, January 2007. https://fas
    .org/sgp/crs/natsec/RL33432.pdf

59 Report: Global drug trafficking market worth half a trillion dollars.
    Talking Drugs, 2017. www.talkingdrugs.org/report-global-illegal-
    drug-trade-valued-at-around-half-a-trillion-dollars

60 UN World Drug Report 2019: 35 million people worldwide suffer
    from drug use disorders while only 1 in 7 people receive
    treatment, 26 June 2019. www.unodc.org/unodc/en/frontpage/
    2019/June/world-drug-report-2019_-35-million-people-worldwide-
    suffer-from-drug-use-disorders-while-only-1-in-7-people-receive-
    treatment.html

Chapter 6: Sick Society

1 Taylor M, Lead poisoning of Port Pirie children: A long history of
   looking the other way. *The Conversation*, 19 July 2012. https://
   theconversation.com/lead-poisoning-of-port-pirie-children-a-long-
   history-of-looking-the-other-way-8296

2 Baghurst PA et al., Environmental exposure to lead and children's
   intelligence at the age of seven years – the Port Pirie cohort study.
   New England Journal of Medicine, 327 (1992), 1279–84. https://pubmed
   .ncbi.nlm.nih.gov/1383818/

3 Landrigan P et al., Pollution and children's health. *Science of The Total
   Environment*, 10 February 2019. www.sciencedirect.com/science/
   article/pii/S0048969718338543

4 Popvich N et al., The Trump Administration is reversing
   100 environmental rules. Here's the full list. *New York Times*, 20 May
   2020. www.nytimes.com/interactive/2020/climate/trump-
   environment-rollbacks.html

5 See www.nytimes.com/2019/08/27/world/americas/bolsonaro-brazil-
   environment.html and www.smh.com.au/politics/federal/fears-
   over-morrison-s-plan-for-single-touch-environmental-approvals-
   20200615-p552uf.html

6   Flynn JR, Massive IQ gains in 14 nations: What IQ tests really
    measure. *Psychological Bulletin*, 1987. https://doi.apa.org/doiLanding?
    doi=10.1037%2F0033–2909.101.2.171

7   Bratsberg B and Rogeberg O, Flynn effect and its reversal are both
    environmentally caused. *PNAS*, 26 June 2018. www.pnas.org/
    content/115/26/6674

8   European Commission, Study for the strategy for a non-toxic
    environment of the 7th Environment Action Programme,
    August 2017.

9   Needleman HL, History of lead poisoning in the world, 1999. www
    .biologicaldiversity.org/campaigns/get_the_lead_out/pdfs/health/
    Needleman_1999.pdf

10 Atkin E, Did Flint's water crisis damage kids' brains? *New Republic*,
    14 February 2018. https://newrepublic.com/article/147066/flints-
    water-crisis-damage-kids-brains

11 China to relocate 15,000 from lead-poisoned area. Agence-France
    Presse, 16 October 2009. https://web.archive.org/web/
    20091019040748/http://news.yahoo.com/s/afp/20091016/hl_afp/
    healthchinaenvironmentpollutionlead

12 Gaylord A et al., Trends in neurodevelopmental disability burden
    due to early life chemical exposure in the USA from 2001 to 2016:
    A population-based disease burden and cost analysis. *Molecular and
    Cellular Endocrinology*, 502 (15 February 2020), 110666. www
    .sciencedirect.com/science/article/abs/pii/S0303720719303685?via%
    3Dihub

13 Criminal Justice, Intelligence and Crime, 2020. https://criminal-
    justice.iresearchnet.com/crime/intelligence-and-crime/3/

14 Bellinger DC, A strategy for comparing the contributions of
    environmental chemicals and other risk factors to
    neurodevelopment of children. *Environmental Health Perspectives*,
    19 December 2011. https://pubmed.ncbi.nlm.nih.gov/22182676/

15 Chen A and Hessler W, Chemical exposures cause child IQ losses
    that rival major diseases. *Environmental Health News*, 24
    February 2012.

16 Cadwalladr C and Graham-Harrison E, Revealed: 50 million
    Facebook profiles harvested for Cambridge Analytica in major data
    breach. *Guardian*, 18 March 2018. www.theguardian.com/news/
    2018/mar/17/cambridge-analytica-facebook-influence-us-election

17 Xia C et al., Lateral orbitofrontal cortex links social impressions to political choices. *Journal of Neuroscience*, 3 June 2015. https://pubmed .ncbi.nlm.nih.gov/26041918/

18 Forrest A, Democracy undergoing 'alarming' decline around the world, study finds. *Independent*, 5 February 2019. www.independent .co.uk/news/world/democracy-freedom-house-annual-report-civil-liberties-authoritarian-donald-trump-us-a8763196.html

19 Forrest A, op. cit.

20 Grandjean P and Landrigan PJ, Developmental neurotoxicity of industrial chemicals: A silent pandemic. *The Lancet*, 368:9353 (2006), 2167–78. https://pubmed.ncbi.nlm.nih.gov/17174709/

21 Toxic exposure: Brain injury. www.braininjury.com/toxic-exposures.shtml#:~:text=There%20are%20nearly%201000% 20substances,cause%20neurological%20and%20brain% 20problems

22 Northstone K et al., Are dietary patterns in childhood associated with IQ at 8 years of age? A population-based cohort study. *Journal of Epidemiology & Community Health*, 66:7 (2011), 624–8.

23 Grandjean P and Landrigan P, op. cit.

24 Chatham-Stephens K et al., Burden of disease from toxic waste sites in India, Indonesia, and the Philippines in 2010. *Environmental Health Perspectives*, 1 July 2013. https://ehp.niehs.nih.gov/doi/10.1289/ehp .1206127

25 See, for example: Jones G and Schneider WJ, Intelligence, human capital, and economic growth: A Bayesian Averaging of Classical Estimates (BACE) approach. *Journal of Economic Growth*, 11:1 (2006), 71–93; Jones G, National IQ and national productivity: The hive mind across Asia. *Asian Development Review*, June 2011; Burhan NA, The impact of low, average, and high IQ on economic growth and technological progress: Do all individuals contribute equally? *Intelligence*, September–October 2014. Also Jones G, *Hive Mind*, Stanford University Press, 2015.

26 Endocrine Society Task Force, Endocrine-disrupting chemicals: An Endocrine Society scientific statement, 2009. https://academic.oup .com/edrv/article/30/4/293/2355049

27 Endocrine-Disrupting Chemicals, An Endocrine Society position statement, 1 May 2018. www.endocrine.org/advocacy/position-statements/endocrine-disrupting-chemicals

28 European Environment Agency, Increase in cancers and fertility problems may be caused by household chemicals and pharmaceuticals, 10 May 2012. www.eea.europa.eu/media/ newsreleases/increase-in-cancers-and-fertility

29 The Endocrine Disruption Exchange, 2020. https:// endocrinedisruption.org/interactive-tools/tedx-list-of-potential- endocrine-disruptors/search-the-tedx-list

30 National Institute of Environmental Health Sciences, Endocrine disruptors, 22 May 2020. www.niehs.nih.gov/health/topics/agents/ endocrine

31 McKie R, £30bn bill to purify water system after toxic impact of contraceptive pill, *Guardian*, 3 June 2012. www.theguardian.com/ environment/2012/jun/02/water-system-toxic-contraceptive-pill#:~: text=Britain%20faces%20a%20%C2%A330bn,European%20Union %20water%20framework%20directive

32 Kolpin DW et al., Pharmaceuticals, hormones, and other organic wastewater contaminants in U.S. streams, 1999–2000: A national reconnaissance. *Environmental Science and Technology*, 36:6 (15 March 2002), 1202–11. https://pubmed.ncbi.nlm.nih.gov/11944670/

33 Harris CA et al., The consequences of feminization in breeding groups of wild fish. *Environmental Health Perspectives*, March 2011. www.ncbi.nlm.nih.gov/pmc/articles/PMC3059991/

34 European Environmental Agency, The impacts of endocrine disruption on wildlife, people and their environments. Technical report No 2/2012. www.eea.europa.eu/publications/the-impacts-of- endocrine-disrupters

35 Swan S in *The Observer*, Mar 28, 2021.

36 Lee JR, Estrogen dominance – an elevated estradiol to progesterone ratio. www.johnleemd.com/estrogen-dominance.html

37 Colborn T, The fossil fuel connection. The Endocrine Disruption Exchange, 2010. https://www.youtube.com/watch?v= 7A7WwRUAB7U

38 Bergman A et al., State of the science of endocrine disrupting chemicals 2012, WHO/UNEP, Geneva, 2013.

39 European Environment Agency, The impacts of endocrine disrupters on wildlife, people and their environments. The Weybridge+15 (1996–2011) report, EEA 2012. https:// www.eea.europa.eu/publications/the-impacts-of-endocrine- disrupters

40 WHO, Global assessment of the state-of-the-science of endocrine disruptors. International Programme on Chemical Safety, WHO/UNEP 2002. https://www.who.int/ipcs/publications/new_issues/endocrine_disruptors/en/

41 Ma Y et al., Effects of environmental contaminants on fertility and reproductive health. *Journal of Environmental Sciences*, 77 (March 2019), 210–17. https://www.sciencedirect.com/science/article/abs/pii/S1001074218313007

42 Horel S and Foucart S, Perturbateurs endocriniens: ces experts contestés qui jouent les semeurs de doute. *Le Monde*, 23 June 2020. www.lemonde.fr/sciences/article/2020/06/22/perturbateurs-endocriniens-ces-experts-contestes-qui-jouent-les-semeurs-de-doute_6043780_1650684.html also (English) www.ehn.org/european-parliament-endocrine-disruptors-2646227143.html

43 Schafer KS and Marquez EC, A generation in jeopardy: How pesticides are undermining our children's intelligence and health. Pesticide Action Network North America, October 2012. http://kresge.org/sites/default/files/Pesticides-childrens-health.pdf

44 PAN International, 2020. www.panna.org/pan-international

45 Boyle CA et al., Trends in the prevalence of developmental disabilities in US children, 1997–2008. *Pediatrics*, 127:6 (2011), 1034–42. https://pubmed.ncbi.nlm.nih.gov/21606152/

46 Perrin JM et al., The rise in chronic conditions among infants, children, and youth can be met with continued health system innovations. Health Affairs, December 2014. www.healthaffairs.org/doi/pdf/10.1377/hlthaff.2014.0832

47 What's behind the stark rise in children's disabilities? NPR, 10 August 2014. www.npr.org/2014/08/19/341674577/whats-behind-the-stark-rise-in-childrens-disabilities

48 Zablotsky B et al., Prevalence and trends of developmental disabilities among children in the U.S: 2009–2017. *Pediatrics*, 26 September 2019. https://doi.org/10.1542/peds.2019-0811

49 Landrigan PJ and Forman J, Chemicals and children's health: The early and delayed consequences of early exposures. Paper presented at WHO Forum, Budapest, 2006. https://www.who.int/ceh/capacity/chemicals.pdf

50 Thomas A et al., Prevalence of attention-deficit/hyperactivity disorder: A systematic review and meta-analysis. *Pediatrics*, April 2015. https://pediatrics.aappublications.org/content/135/4/e994.short

51 American Psychiatric Association. What is ADHD? APA website, July 2017. www.psychiatry.org/patients-families/adhd/what-is-adhd

52 ADHD Statistics: How common is ADHD? May 2020. www .additudemag.com/statistics-of-adhd/

53 WHO, Autism spectrum disorders, November 2019. https://www .who.int/news-room/fact-sheets/detail/autism-spectrum-disorders

54 CDC, Data & statistics on autism spectrum disorder, 2020. www.cdc .gov/ncbddd/autism/data.html

55 Grandjean P and Landrigan PJ, Developmental neurotoxicity of industrial chemicals: A silent pandemic. *The Lancet*, 368:9353 (2006), 2167–78. https://pubmed.ncbi.nlm.nih.gov/17174709/

56 Grandjean P and Landrigan P, Neurobehavioural effects of developmental toxicity. *Lancet Neurology*, 13:3 (May 2015), 330–8. www.ncbi.nlm.nih.gov/pmc/articles/PMC4418502/

57 Volk H et al., Residential proximity to freeways and autism in the CHARGE study. *Environmental Health Perspectives*, 119:6 (2011), 873–7. www.ncbi.nlm.nih.gov/pmc/articles/PMC3114825

58 Weisskopf M et al., Perinatal air pollution exposure and autism, with new results in the Nurses' Health Study. International Society for Autism Research, August 2013. https://pubmed.ncbi.nlm.nih .gov/23816781/

59 WHO, Air pollution, 2020. www.who.int/health-topics/air-pollution#tab=tab_1

60 WHO, Cancer in children, 2020. www.who.int/news-room/fact-sheets/detail/cancer-in-children

61 Van Larebeke NA et al., Unrecognized or potential risk factors for childhood cancer. *International Journal of Occupational Environment Health*, 11:2 (2005), 199–201. https://pubmed.ncbi.nlm.nih.gov/15875896/

62 Lavigne E et al., Maternal exposure to ambient air pollution and risk of early childhood cancers: A population-based study in Ontario, Canada. *Environment International*, March 2017. https://pubmed.ncbi .nlm.nih.gov/28108116/

63 Youlden DR et al., The incidence of childhood cancer in Australia, 1983–2015, and projections to 2035. *Medical Journal of Australia*, 26 December 2019. https://www.mja.com.au/journal/2020/212/3/incidence-childhood-cancer-australia-1983-2015-and-projections-2035

64 WHO, Asthma, 2020. https://www.who.int/news-room/fact-sheets/detail/asthma

65 Khreis H et al., Full-chain health impact assessment of traffic-related air pollution and childhood asthma. *Environment International*, May 2018. https://www.sciencedirect.com/science/article/pii/S0160412017320184; Alcock I et al., Land cover and air pollution are associated with asthma hospitalisations: A cross-sectional study. *Environment International*, December 2017. https://www.sciencedirect.com/science/article/pii/S0160412017304026

66 Yuexia S et al., Modern life makes children allergic. A cross-sectional study: Associations of home environment and lifestyles with asthma and allergy among children in Tianjin region, China. *International Archives of Occupational and Environmental Health*, 9 January 2019. https://pubmed.ncbi.nlm.nih.gov/30627853/

67 Asthma and Allergy Foundation of America, Asthma facts and figures. AAFA 2020. www.aafa.org/asthma-facts/

68 CDC, Asthma as the underlying cause of death https://www.cdc.gov/asthma/asthma_stats/asthma_underlying_death.html#:~:text=Adults%20were%20nearly%20five%20times,with%20all%20other%20age%20groups, 24 April 2018.

69 UK National Health Service, Medically unexplained symptoms. www.nhs.uk/conditions/medically-unexplained-symptoms/

70 Kirkmayer LJ et al., Explaining medically unexplained symptoms. *Canadian Journal of Psychiatry*, 49:10 (2004), 663–72. https://pubmed.ncbi.nlm.nih.gov/15560312/

71 Mayo Clinic, Morgellons disease: Managing a mysterious skin condition, 2 April 2020. www.mayoclinic.com/health/morgellons-disease/SN00043.

72 US Department of Veterans Affairs, Veterans' diseases associated with Agent Orange, 2020. www.publichealth.va.gov/exposures/agentorange/conditions/index.asp

73 Wang A et al., Parkinson's disease risk from ambient exposure to pesticides. *European Journal of Epidemiology*, 26:7 (2011), 547–55. https://pubmed.ncbi.nlm.nih.gov/21505849/

74 Hayden K et al., Occupational exposure to pesticides increases the risk of incident AD: The Cache County study. *Neurology*, 74:19 (2010), 1524–30. https://pubmed.ncbi.nlm.nih.gov/20458069/

75 WHO, Dementia, 2019. https://www.who.int/news-room/fact-sheets/detail/dementia

76 Burger L and Bellon T, Bayer to pay up to $10.9 billion to settle bulk of Roundup weedkiller cancer lawsuits. Reuters, 25 June 2020. www.reuters.com/article/us-bayer-litigation-settlement-idUSKBN23V2NP

77 Gillam C, I won a historic lawsuit, but may not live to get the money. *Time* magazine, 21 November 2018. https://time.com/5460793/dewayne-lee-johnson-monsanto-lawsuit/

78 Hakim D, Monsanto emails raise issue of influencing research on Roundup weed killer. *New York Times*, 1 August 2017. https://www.nytimes.com/2017/08/01/business/monsantos-sway-over-research-is-seen-in-disclosed-emails.html

79 IARC, IARC Monograph on glyphosate, March 2015. www.iarc.fr/wp-content/uploads/2018/07/MonographVolume112-1.pdf

80 EPA, EPA finalizes glyphosate mitigation, 30 January 2020. https://www.epa.gov/pesticides/epa-finalizes-glyphosate-mitigation

81 Hessler U, What's driving Europe's stance on glyphosate. MSN News, 26 June 2020. www.msn.com/en-za/news/other/whats-driving-europes-stance-on-glyphosate/ar-BB15XlQZ

82 BSI, Are you #AutoImmuneAware? 2018. https://www.immunology.org/news/new-connect-immune-research-report-are-you-autoimmuneaware

83 Pollard KM, Environment, autoantibodies, and autoimmunity. *Frontiers in Immunology*, 11 February 2015. https://www.frontiersin.org/articles/10.3389/fimmu.2015.00060/full

84 Globocan, All cancers fact sheet, WHO 2018. https://gco.iarc.fr/today/data/factsheets/cancers/39-All-cancers-fact-sheet.pdf

85 IARC, Cancer Tomorrow, WHO 2020. https://gco.iarc.fr/tomorrow/home

86 World Cancer Research Fund, Global cancer data per country, 2018. https://www.wcrf.org/dietandcancer/cancer-trends/data-cancer-frequency-country

87 Wild C et al., World Cancer Report 2014. International Agency for Research on Cancer (IARC). www.esmo.org/oncology-news/world-cancer-report-2014

88 IARC, IARC Monographs on the identification of carcinogenic hazards to humans. WHO/IARC, 2020. https://monographs.iarc.fr/agents-classified-by-the-iarc/

89 IARC, Classifications by cancer site, 2020. https://monographs.iarc.fr/wp-content/uploads/2019/07/Classifications_by_cancer_site.pdf

90 Agency for Toxic Substances and Disease Registry, Cancer, chemicals and you. https://www.atsdr.cdc.gov/emes/public/docs/Chemicals,%20Cancer,%20and%20You%20FS.pdf

91 National Cancer Institute, The genetics of cancer, 2020. https://
    www.cancer.gov/about-cancer/causes-prevention/genetics

92 WHO, Cancer prevention, 2020. https://www.who.int/health-
    topics/cancer#tab=tab_2

93 Germanos A, New report urges research into environmental
    causes of breast cancer. Common Dreams, 12 February 2013.
    https://www.commondreams.org/news/2013/02/12/new-report-
    urges-research-environmental-causes-breast-cancer

94 Forman MR et al., Breast cancer and the environment: Prioritizing
    prevention. US Interagency Breast Cancer and Environmental
    Research Coordinating Committee, February 2013. https://www
    .niehs.nih.gov/about/assets/docs/breast_cancer_and_the_
    environment_prioritizing_prevention_508.pdf

95 Fenga C, Occupational exposure and risk of breast cancer.
    Biomedical Reports, March 2016. https://pubmed.ncbi.nlm.nih.gov/
    26998264/

96 Shepherd N and Parker C, Depression in adults: Recognition and
    management. *The Pharmaceutical Journal*, 5 April 2017. https://www
    .pharmaceutical-journal.com/cpd-and-learning/cpd-article/
    depression-in-adults-recognition-and-management/20202439
    .cpdarticle?firstPass=false and WHO, Depression, 30 January 2020.
    https://www.who.int/news-room/fact-sheets/detail/depression

97 Jowitt J, What is depression and why is it rising? *Guardian*, 4 June
    2018. www.theguardian.com/news/2018/jun/04/what-is-
    depression-and-why-is-it-rising

98 Pies RW, Debunking the two chemical imbalance myths, again.
    Psychiatric Times, 2 August 2019. https://www.psychiatrictimes
    .com/view/debunking-two-chemical-imbalance-myths-again

99 Harvard Medical School, What causes depression? 24 June 1019.
    www.health.harvard.edu/mind-and-mood/what-causes-depression

100 van den Bosch M and Meyer-Lindenberg A, Environmental
    exposures and depression: Biological mechanisms and
    epidemiological evidence. *Annual Review of Public Health*, 2019.
    www.annualreviews.org/doi/pdf/10.1146/annurev-publhealth-
    040218-044106

101 Stallones L and Beseler C, A cohort study of pesticide poisoning
    and depression in Colorado farm residents. *Annals of Epidemiology*,
    12:6 (May 2008). https://pubmed.ncbi.nlm.nih.gov/18693039/

102 Rehner TA et al., Depression among victims of South Mississippi's
     methyl parathion disaster. *Health & Social Work*, February 2000.
     https://pubmed.ncbi.nlm.nih.gov/10689601/

103 Kim J, Ko Y and Lee WJ. Depressive symptoms and severity of
     acute occupational pesticide poisoning among male farmers.
     Occupational and Environmental Medicine, 70:5 (2013), 303–9. https://
     pubmed.ncbi.nlm.nih.gov/23390200/

104 Yazd SD et al., Key risk factors affecting farmers' mental health:
     A systematic review. *International Journal of Environmental Research
     and Public Health*, 2 December 2019. https://pubmed.ncbi.nlm.nih
     .gov/31810320/

105 Breggin PR, Antidepressants can cause long-term depression.
     Huffington Post, 16 November 2011. https://www.huffpost.com/
     entry/antidepressants-long-term-depression_b_1077185

106 WHO, Obesity and overweight, April 2020. www.who.int/news-
     room/fact-sheets/detail/obesity-and-overweight

107 WHO, Diabetes, 2020. www.who.int/health-topics/diabetes#tab=tab_1

108 Porta M and Lee DH, Review of the science linking chemical
     exposures to the human risk of obesity and diabetes. ChemTrust
     UK, March 2012. www.wecf.eu/download/2012/March/
     CHEMTrustObesityDiabetesSummaryReport.pdf

109 Braun J, Early life exposure to endocrine disrupting chemicals and
     childhood obesity and neurodevelopment. *Nature Reviews
     Endocrinology*, 13:3 (March 2017), 161–73. www.ncbi.nlm.nih.gov/
     pmc/articles/PMC5322271/

110 Egusquiza RJ and Blumberg B, Environmental obesogens and their
     impact on susceptibility to obesity: New mechanisms and
     chemicals. *Endocrinology*, 161:3 (March 2020). https://academic.oup
     .com/endo/article/161/3/bqaa024/5739626

111 Holliday R, Epigenetics: A historical overview, *Epigenetics*, 1:2
     (March 2006). www.tandfonline.com/doi/abs/10.4161/epi.1.2
     .2762

112 Food addiction and the brain. The Health Report, ABC Radio
     National, 5 August 2013. https://www.abc.net.au/radionational/
     programs/healthreport/food-addiction/4865260

113 Hou L et al., Environmental chemical exposures and human
     epigenetics. *International Journal of Epidemiology*, 41:1 (2012),
     79–105. https://doi.org/10.1093/ije/dyr154

114 Marczylo EL et al., Environmentally induced epigenetic toxicity: Potential public health concerns. *Critical Reviews in Toxicology*, 13 September 2016. https://www.ncbi.nlm.nih.gov/pmc/articles/PMC5030620/

115 Roser M and Ritchie H, Burden of disease. Our World in Data, 2020. https://ourworldindata.org/burden-of-disease

116 Poulos A, The secret life of chemicals: A guide to chemicals in our environment and how to protect yourself, 2019. https://professoralfredpoulos.com/the-secret-life-of-chemicals/

Chapter 7: Getting Away with Murder

1 Smith WE and Smith AM, *Minamata: The Story of the Poisoning of a City, and of the People Who Choose to Carry the Burden of Courage*, Holt, Reinhart and Winston, 1975.

2 Peterson MJ, Case study: Bhopal plant disaster. International Dimensions of Ethics Education Case Study Series, 20 March 2009. https://scholarworks.umass.edu/edethicsinscience/4/

3 Amnesty International, Clouds of injustice: Bhopal disaster 20 years on, 2004. https://www.amnesty.org/en/documents/ASA20/015/2004/en/

4 Ellis-Petersen H, Bhopal's tragedy has not stopped: The urban disaster still claiming lives 35 years on. *Guardian*, 8 December 2019. www.theguardian.com/cities/2019/dec/08/bhopals-tragedy-has-not-stopped-the-urban-disaster-still-claiming-lives-35-years-on

5 Ellis-Petersen H, 2019, op. cit.

6 Griffin P, Carbon majors report, Climate Accountability Institute, July 2017. https://climateaccountability.org/carbonmajors.html

7 Big oil's real agenda on climate change, InfluenceMap, March 2019. https://influencemap.org/report/How-Big-Oil-Continues-to-Oppose-the-Paris-Agreement-38212275958aa21196dae3b76220bddc

8 Holden E, How the oil industry has spent billions to control the climate change conversation. *Guardian*, 9 January 2020. www.theguardian.com/business/2020/jan/08/oil-companies-climate-crisis-pr-spending

9 Poulos A, 2019, op. cit.

10 Stockholm Convention, 2020. www.pops.int/TheConvention/
    ThePOPs/The12InitialPOPs/tabid/296/Default.aspx and www.pops
    .int/TheConvention/ThePOPs/TheNewPOPs/tabid/2511/Default.aspx

11 UNEP Global Chemical Outlook, 2013, op. cit.

12 Milman O, US cosmetics are full of chemicals banned by Europe –
    why? *Guardian*, 22 May 2019. www.theguardian.com/us-news/2019/
    may/22/chemicals-in-cosmetics-us-restricted-eu

13 UNEP, 2019, op. cit.

14 UNEP, 2013, op. cit.

15 Cohen J. Chemical time bomb. Four Corners, ABC TV, 22 July 2013.
    www.abc.net.au/4corners/chemical-time-bomb/4836466

16 US FDA, Import refusals. www.accessdata.fda.gov/scripts/
    ImportRefusals/index.cfm

17 Popovich N et al., The Trump Administration is reversing
    100 environmental rules. Here's the full list. *New York Times*, 20 May
    2020. www.nytimes.com/interactive/2020/climate/trump-
    environment-rollbacks.html

18 Cutler D and Dominici F, A breath of bad air: Cost of the Trump
    environmental agenda may lead to 80 000 extra deaths per decade.
    JAMA, 319:22 (2018), 2261–2. https://jamanetwork.com/journals/
    jama/fullarticle/2684596?appid=scweb&appid=scweb&alert=article

19 Hong S et al., *China's chemical industry: New strategies for a new era.*
    McKinsey, 20 March 2019. www.mckinsey.com/industries/
    chemicals/our-insights/chinas-chemical-industry-new-strategies-for-
    a-new-era#

20 American Chemistry Association, Responsible care by the numbers,
    August 2019. https://responsiblecare.americanchemistry.com/
    FactSheet/

21 Responsible Care, Guiding principles, ACA, 2020. https://
    responsiblecare.americanchemistry.com/ResponsibleCare/
    Responsible-Care-Program-Elements/Guiding-Principles/

22 Royal Society of Chemistry, Regulation of the Profession and Code
    of Conduct, 2016. www.rsc.org/globalassets/03-membership-
    community/join-us/membership-regulations/code-of-conduct.pdf

23 Royal Australian Chemical Institute, Code of Ethics, 2014.

24 Rosenberg D, Profile of the chemical industry: Power, corruption &
    lies. National Resources Defense Council (NRDC), 11 May 2012.
    www.nrdc.org/experts/daniel-rosenberg/profile-chemical-industry-
    power-corruption-lies

25 UNEP, Global Chemicals Outlook, 2013.
26 Fast NJ, How to stop the blame game. *Harvard Business Review*,13 May 2010. https://hbr.org/2010/05/how-to-stop-the-blame-game

Chapter 8: Clean Up Society

1 Grandjean P and Landrigen PJ, Developmental neurotoxicity of industrial chemicals. *The Lancet*, 8 November 2006. https://www.thelancet.com/journals/lancet/article/PIIS0140-6736(06)69665-7/fulltext
2 Haynes J, Socio-economic impact of the Sydney 2000 Olympic Games. Centre d'Estudis Olímpics UAB, Barcelona, 2001.
3 Sydney Olympic Park Authority, Site remediation, 2011.
4 Hunt J et al., Homebush Bay sediment remediation: A case study. Thiess Services Pty Ltd, Sydney, 2009.
5 Naidu R, *Contamination: Big Risks, Bigger Opportunities*. CRC CARE, Salisbury, Australia, 2012.
6 Loong LH, Clean and Green Singapore 2014 launch speech. https://www.pmo.gov.sg/Newsroom/speech-prime-minister-lee-hsien-loong-launch-clean-and-green-singapore-2014
7 Singapore Government, Zero Waste Master Plan, 2020. https://www.towardszerowaste.gov.sg/
8 Singapore Government, Our Throwaway Culture, 2020. www.towardszerowaste.gov.sg/zero-waste-masterplan/chapter1/case-for-zero-waste
9 Singapore Government, Clean and Green Singapore, 2020. www.nea.gov.sg/programmes-grants/campaigns/clean-and-green-singapore
10 US EPA, National Priorities List (NPL) Sites – by State, 2020. https://www.epa.gov/superfund/national-priorities-list-npl-sites-state
11 European Commission, European achievements in soil remediation and brownfield redevelopment, 2017. https://ec.europa.eu/jrc/en/publication/european-achievements-soil-remediation-and-brownfield-redevelopment
12 Li X et al., Contaminated sites in China: Countermeasures of provincial governments. *Journal of Cleaner Production*, March 2017. www.sciencedirect.com/science/article/abs/pii/S0959652617301221
13 Dong Z et al., Financing models for soil remediation in China. IISD, 2018. www.iisd.org/library/financing-models-soil-remediation-china

14 Xu F et al., Tianjin chemical clean-up after explosion, 22 September 2015. www.cmaj.ca/content/187/13/E404#:~:text=A%20massive% 20clean-up%20of%20contaminated%20water%20is%20underway, more%20than%20700%3B%2069%20people%20are%20still% 20missing

15 UNEP, Global Chemicals Outlook II, 2019, op. cit.

16 UNEP, Global Chemicals Outlook II, 2019, op. cit.

17 Guinee JB et al., Life cycle assessment: Past, present, and future. *Environmental Science and Technology*, 45:1 (2011), 90–6. https://pubs .acs.org/doi/10.1021/es101316v

18 Brunner PH and Rechberger H, *Practical Handbook of Material Flow Analysis*, Lewis, 2016.

19 Janse B, Multiple criteria decision analysis (MCDA), 26 August 2018. www.toolshero.com/decision-making/multiple-criteria-decision- analysis-mcda/

20 See, for example, Mitsui Chemicals Group, Product Stewardship by the Mitsui Chemicals Group, 2020. https://jp.mitsuichemicals.com/ en/sustainability/rc/chemicals/

21 OECD, Extended producer responsibility, 2020. http://www.oecd .org/environment/extended-producer-responsibility.htm

22 US EPA, Safer chemical ingredients list, 2020. https://www.epa.gov/ saferchoice/safer-ingredients

23 American Chemical Society, What is green chemistry? 2020. https:// www.acs.org/content/acs/en/greenchemistry/what-is-green- chemistry.html

24 Tricoire JP. Here's why green manufacturing is crucial for a low- carbon future. World Economic Forum, 22 January 2019. https:// www.weforum.org/agenda/2019/01/here-s-why-green- manufacturing-is-crucial-for-a-low-carbon-future/

25 World Green Building Council, About green building, 2020. https:// www.worldgbc.org/what-green-building

26 US EPA, Integrated pest management (IPM) principles, 2020. https:// www.epa.gov/safepestcontrol/integrated-pest-management-ipm- principles

27 Terra Genesis International, Regenerative agriculture, 2020. www .regenerativeagriculturedefinition.com/

28 UNIDO, What is chemical leasing? 2020. www.chemicalleasing.org/ what-chemical-leasing

29 Frosch RA and Gallopoulos NE, Strategies for manufacturing. *Scientific American*, 1989. www.jstor.org/stable/24987406?seq=1

30 Zaman AU. A comprehensive review of the development of zero waste management: Lessons learned and guidelines. *Journal of Cleaner Production*, 91 (15 March 2015), 12–25. https://www.sciencedirect.com/science/article/abs/pii/S0959652614013018

31 EPA of South Australia, Guidelines for the assessment and remediation of site contamination, November 2019. www.epa.sa.gov.au/files/13544_sc_groundwater_assessment.pdf

32 Siow JCF, *Health Impacts on Persistent Organic Pollutants and/or Heavy Metals*, CRC CARE, Salisbury, Australia, 2013.

33 See, for example, Jelinek G, The influence of the pharmaceutical industry in medicine. *Journal of Law and Medicine*, 17:2 (October 2009), 216–23. www.researchgate.net/publication/40646831_The_influence_of_the_pharmaceutical_industry_in_medicine

34 Vigdor N, It paid doctors kickbacks. Now, Novartis will pay a $678 million settlement. *New York Times*, 1 July 2020. www.nytimes.com/2020/07/01/business/Novartis-kickbacks-diabetes-heart-drugs.html

35 Moynihan R and Henry D, The fight against disease mongering: Generating knowledge for action, *PLoS Med*, 3:4 (2006). https://journals.plos.org/plosmedicine/article?id=10.1371/journal.pmed.0030191

36 E.g. Egilman DS and Bohme SR, Over a barrel: Corporate corruption of science and its effects on workers and the environment. *International Journal of Occupational and Environmental Health*, October 2005; Heads they win, tails we lose: How corporations corrupt science at the public's expense. Union of Concerned Scientists, 17 February 2012; The climate denial machine: How the fossil fuel industry blocks climate action. The Climate Reality Project, 5 September 2019.

37 US Congress, Letter to USEPA Administrator Scott Pruitt, 21 May 2018. https://energycommerce.house.gov/sites/democrats.energycommerce.house.gov/files/documents/EPA.2018.05.21.%20Letter%20re%20ACC%20and%20PFAS.%20EE.OI_.PDF

38 CRC CARE, PFAS contamination, 2020. https://www.crccare.com/knowledge-sharing/pfas-contamination

39 Light DW, New prescription drugs: A major health risk with few offsetting advantages. Harvard University Center for Ethics, 2014. https://ethics.harvard.edu/blog/new-prescription-drugs-major-health-risk-few-offsetting-advantages

40 Wang L, The struggle to keep women in academia. *Chemistry and Engineering News*, 12 May 2019. https://cen.acs.org/careers/diversity/struggle-keep-women-academia/97/i19

41 Tullo AH, Women in the chemical industry 2017. *Chemistry and Engineering News*, 11 September 2017. https://cen.acs.org/articles/95/i36/Women-chemical-industry-2017.html

42 A Chemical Imbalance. www.chemicalimbalance.co.uk/project/watch-the-film/

43 Yamas V et al. The young male syndrome: An analysis of sex, age, risk taking and mortality in patients with severe traumatic brain injuries. *Frontiers of Neurology*, 12 April 2019. https://www.frontiersin.org/articles/10.3389/fneur.2019.00366/full#B1

44 For why the world needs female leadership in order to survive, see Cribb JHJ, *Surviving the 21st Century*, Springer, 2017.

45 Holden C and Mace R, Phylogenetic analysis of the evolution of lactose digestion in adults. *Human Biology*, 1997. www.jstor.org/stable/41466630?seq=1

46 Beall CM, Andean, Tibetan, and Ethiopian patterns of adaptation to high-altitude hypoxia. *Integrative and Comparative Biology*, February 2006. https://academic.oup.com/icb/article/46/1/18/661204

47 Schlebusch CM et al., Human adaptation to arsenic-rich environments. *Molecular Biology and Evolution*, 32:6 (June 2015). https://academic.oup.com/mbe/article/32/6/1544/1074042

48 International Energy Agency, Global Energy Review, 2020. https://www.iea.org/reports/global-energy-review-2020/renewables#abstract

49 Zhang L and Xu Z. A review of current progress of recycling technologies for metals from waste electrical and electronic equipment. *Journal of Cleaner Production*, 20 July 2016. www.sciencedirect.com/science/article/abs/pii/S0959652616302451

50 Nace, T. World population expected to peak in just 44 years as fertility rates sink, Forbes, 17 July 2020. https://www.forbes.com/sites/trevornace/2020/07/17/world-population-expected-to-peak-in-just-44-years-as-fertility-rates-sink/?sh=42f1bf8b372a

51 UN, Growing at a slower pace, world population is expected to reach 9.7 billion in 2050 and could peak at nearly 11 billion around 2100, 17 June 2019. https://www.un.org/development/desa/en/news/population/world-population-prospects-2019.html

52 Joardder MUH et al., Solar pyrolysis: Converting waste into asset using solar energy. In *Clean Energy for Sustainable Development*, Elsevier, 2017.

53 Cribb JHJ, *Food or War*, Cambridge University Press, 2019.

54 Tuomisto HI and Texeira de Mattos MJ, Environmental impacts of cultured meat production. *Environmental Science and Technology*, 45:14 (17 June 2011), 6117–23. https://doi.org/10.1021/es200130u
55 See Cribb JHJ, op. cit., pp. 243 et seq.

Chapter 9: Clean Up the Earth

1 Chuda N and J, The Founders. Healthy Child Healthy World, 2020. https://healthychild.org/about/founders/
2 Sharp CR et al., Parental exposures to pesticides and risk of Wilms' tumor in Brazil. *American Journal of Epidemiology*, 141:3 (1995), 210–17. www.ncbi.nlm.nih.gov/pubmed/7840094
3 Fear NT et al., Childhood cancer and paternal employment in agriculture: The role of pesticides. *British Journal of Cancer*, March 1998. https://pubmed.ncbi.nlm.nih.gov/9514065/
4 Chuda N and J, 2020, op. cit.
5 Boxer B, in Congressional Record, Proceedings and Debates of the 104th Congress, Second Session. Washington, Thursday, 9 May 1996.
6 Wikipedia, Nestle boycott, 2020. https://en.wikipedia.org/wiki/Nestl%C3%A9_boycott
7 Boycott Nike, 1996. www.saigon.com/nike/
8 Ackerley BA, Creating justice for Bangladeshi garment workers with pressure not boycotts. The Conversation, 25 April 2015. https://theconversation.com/creating-justice-for-bangladeshi-garment-workers-with-pressure-not-boycotts-40592
9 About us, Ethical Consumer, 2020. www.ethicalconsumer.org/about-us
10 History of successful boycotts, Ethical Consumer, 2020. www.ethicalconsumer.org/ethicalcampaigns/boycotts/history-successful-boycotts
11 Beck V, Consumer boycotts as instruments for structural change. *Journal of Applied Philosophy*, 20 January 2018. https://doi.org/10.1111/japp.12301
12 The ultimate guide to corporate social responsibility (CSR), Businesses for Good (Big1.com), 2020. https://www.b1g1.com/businessforgood/csr-guide
13 McDonnell MH and King B, Keeping up appearances: Reputational threat and impression management after social movement boycotts. *Administrative Science Quarterly*, 9 August 2013. https://journals.sagepub.com/doi/abs/10.1177/0001839213500032

14   Cribb JHJ, The Nation State is on the skids. MAHB, 15 December 2015. https://mahb.stanford.edu/blog/nation-state-on-the-skids/

15   Avaaz, The Avaaz way: How we work, 2020. https://secure.avaaz.org/page/en/about/

16   Our Campaigns, SumOfUs, 2020. www.sumofus.org/campaigns/

17   350.org, 350 Celebrates a decade of action, 2019. https://350.org/10-years/

18   https://saferchemicals.org/

19   Greenpeace SEAsia, Q: How did Greenpeace start? 2020. https://www.greenpeace.org.uk/about-greenpeace/

20   Terras P, 17,000 hazardous chemicals and counting – open-sourcing Greenpeace's global chemical research. Greenpeace, 2 April 2016. www.greenpeace.org/international/story/7526/17000-hazardous-chemicals-and-counting-open-sourcing-greenpeaces-global-chemical-research/

21   Pesticides, waterways & drinking water, FoEA, March 2020. https://www.foe.org.au/pesticides

22   PAN Pesticides Database – Chemicals, Pesticide Action Network, 2020. http://www.panna.org/legacy/panups/panna-new-pan-pesticide-database

23   Circle of Blue. www.circleofblue.org/about/

24   Threats: Pollution, WWF, 2020. https://www.worldwildlife.org/threats/pollution

25   De Chardin T. *The Future of Man*, 1946. English edition, Harper & Row, 1964.

26   Humans are 'learning to think as a species. SciNews, PhysOrg, 28 March 2017. https://phys.org/news/2017-03-humans-species.html

27   Clement J. Worldwide digital population as of July 2020. Statista, 24 July, 2020. www.statista.com/statistics/617136/digital-population-worldwide/

28   Cribb JHJ, New name needed for unwise Homo? *Nature*, 476 (17 August 2011), 282. www.nature.com/articles/476282b

29   The 2020 World CleanUp conference was cancelled due to the Covid pandemic.

30   Naidu R, The Global Contamination Initiative. CRC CARE, Salisbury, Australia, 2013. https://www.crccare.com/files/dmfile/GlobalcontaminationintiativepresentationfromCleanUp20132.pdf

31   Wang Z et al., Toward a global understanding of chemical pollution: A first comprehensive analysis of national and regional chemical

inventories. *Environmental Science and Technology*, 54:5 (2020), 2575–84. https://pubs.acs.org/doi/abs/10.1021/acs.est.9b06379

32 Hippocratic Oath. https://en.wikipedia.org/wiki/Hippocratic_Oath

33 Smith CM, Origin and uses of *primum non nocere* – Above all, do no harm! *Clinical Pharmacology*, 7 March 2013. https://doi:org/10.1177/0091270004273680

34 In 'practitioners', it is desirable to include business managers, even sales representatives, for chemical companies.

35 In case this should be thought impossible, a number of supercomputer programs are under development to model and predict chemical interactions and potential toxicity.

36 United Nations, 1948. Universal Declaration of Human Rights. https://www.ohchr.org/EN/UDHR/Documents/UDHR_Translations/eng.pdf

Chapter 10: Preventing Catastrophe

1 Cribb JHJ, *Surviving the 21st Century*, Springer, 2017.

2 See Cribb JHJ, *Food or War*, Cambridge University Press, 2019.

3 Mecklin J, Closer than ever: It is 100 seconds to midnight. *Bulletin of the Atomic Scientists*, 23 January 2020. https://thebulletin.org/doomsday-clock/https://thebulletin.org/doomsday-clock/

4 IPCC, Causes of climate change. Climate Change 2014 Synthesis Report. www.ipcc.ch/site/assets/uploads/2018/02/AR5_SYR_FINAL_SPM.pdf

5 Ecological footprint, Global Footprint Network, 2020. https://www.footprintnetwork.org/our-work/ecological-footprint/

6 Cribb JHJ, 2019, op. cit.

7 Bostrom N, Existential risks: Analyzing human extinction scenarios and related hazards. *Journal of Evolution and Technology*, 9:1 (2002). https://nickbostrom.com/existential/risks.html

Postscript

1 Wikipedia, Great Oxidation Event, 2020. https://en.wikipedia.org/wiki/Great_Oxidation_Event; BBC, The event that transformed the Earth, 2015. www.bbc.com/earth/story/20150701-the-origin-of-the-air-we-breathe

INDEX

Locators in **bold** refer to tables; those in *italic* to figures